FRASER VALLEY REGIONAL LIBRARY

39083502358063

Some Like It Cold

D0791710

Some Like It Cold

The Politics of Climate Change in Canada

Robert C. Paehlke

Between the Lines
Toronto

Some Like it Cold: The Politics of Climate Change in Canada

© 2008 by Robert C. Paehlke

First published in 2008 by
Between the Lines
720 Bathurst Street, Suite #404
Toronto, Ontario
M5S 2R4

1-800-718-7201

www.btlbooks.com

All rights reserved. No part of this publication may be photocopied, reproduced, stored in a retrieval system, or transmitted in any form or by any means, electronic, mechanical, recording, or otherwise, without the written permission of Between the Lines, or (for photocopying in Canada only) Access Copyright, 1 Yonge Street, Suite 1900, Toronto, Ontario, M5E 1E5.

Every reasonable effort has been made to identify copyright holders. Between the Lines would be pleased to have any errors or omissions brought to its attention.

Paehlke, Robert
        Some like it cold : the politics of climate change in Canada / Robert C. Paehlke
Includes index.
ISBN 978–1–897071–39–7

1.   1. Climatic changes – Political aspects – Canada.   2. Political participation – Canada.   3. Global warming – Government policy – Canada.   4. Environmental policy – Canada – Citizen participation.   I. Title.
QC981.8.C5P33 2008   363.738′740971   C2008-901331-X

Cover design by Jennifer Tiberio
Front cover image ©iStockphoto.com/Jan Will
Text design and page preparation by Steve Izma
Printed in Canada

Between the Lines gratefully acknowledges assistance for its publishing activities from the Canada Council for the Arts, the Ontario Arts Council, and the Government of Canada through the Book Publishing Industry Development Program.

 Canada Council   Conseil des Arts
for the Arts   du Canada

Canada

 ONTARIO ARTS COUNCIL
CONSEIL DES ARTS DE L'ONTARIO

Contents

Acknowledgements

I WOULD LIKE to thank the collective members of Between the Lines, especially Robert Clarke, for suggesting this project; and Keith Stewart and Jamie Swift for making many useful suggestions for improvement. I would also like to thank Birgitte Berkowitz for her continuing encouragement and support in all things.

One

Introduction

A Personal Reflection

BORN AND RAISED near New York City, I moved to Canada in 1967 as a young man, settling in Peterborough, Ontario, in 1970. I immediately felt comfortable in my new country. Soon I was wondering why a person so thoroughly Canadian in outlook had been born so far to the south. My only not-so-Canadian inclinations were that I found hockey a bit violent and I did not like the enduring cold of Ontario winters.

I remember telling American friends and relatives about how the snow that fell in November was almost always still around in early April. I found this both wondrous and a little scary. I was comforted when I learned that retired Ontarians just fled for Florida for the winter after their Christmas in the true north brave and free (if not sooner).

What I never imagined then was that by the time I retired the relatively mild winters of my New Jersey youth would have migrated 800 kilometres north all the way to Central Ontario. Who needs Florida when for three of the past four years people have been playing golf in Peterborough in December or even January? At first I played too, just to say I had done it, and then I did it because the opportunities for cross-country skiing were becoming less and less frequent.

In 1990 I purchased the house where I still live in part because it was one hundred yards from the entrance to an extensive trail system

1

that runs through a large park and out into the countryside. I still use that trail for walking or cycling, but now if I want to ski more than a few times each winter I need to pack up my skis and poles and get into my car instead of skiing from my doorstep. Since the golf courses are much closer, my environmental conscience sends me out to them in December instead of into my car and north to the Canadian Shield.

The winter of 2006–7 was especially dramatic in its inability to actually be winter. We didn't even need to dress all that warmly to play golf into the New Year. Everyone was talking about the difference and appreciated that the weather was so out of line with the norm that it was less a matter of odd weather than a clear sign of an emerging new climate. That change is now apparent everywhere, especially in more northern settings in the United States, in Canada, and throughout the polar regions of the planet.

In the winter that wasn't, people in New York City were wearing shorts and short-sleeved shirts to do their Christmas shopping. My childhood and young adult memories in New Jersey, Pennsylvania, and New York City are of snowmen, or at least a lot of slush. The climate statistics bear out these impressions, and because winters are always variable we need to know those statistics before we leap to conclusions; but there are many forms of knowing, and with recent winters many people *really knew* in ways they hadn't previously.[1]

The dramatic differences that we witnessed, worried about, and enjoyed in Canada underscored the arrival of Stéphane Dion as leader of the Liberal Party. Dion, a former environment minister, focused his leadership campaign on climate change. Almost as dramatic an event in Canadian politics is the emergence of a politically viable Green Party, and equally so in the United States is the success of Al Gore's film *An Inconvenient Truth*. These events were followed immediately by the dramatic weather of late 2006 and early 2007. It was almost as if Hollywood marketers had suddenly attained godlike powers and Canada's Liberal rainmakers had gone literal and somehow managed to put a stop to the snow that should have fallen the Christmas after Dion was chosen.

For me, winter became less scary, but I am not a polar bear. Nor am

I an African or Australian farmer dealing with deepening drought, nor do I live in a nation that might soon be underwater. Citizens of the latter would include Bangladeshis, selected island dwellers around the world, and people in a United States whose southern shores may one day be somewhere around Arkansas. The benign winter that brought golf to Ontario was also part of an emerging weather pattern that brought devastating El Niño storms that extensively damaged Stanley Park in Vancouver and killed mountaineers and a loving father in Oregon. Moreover, few would deny that the intensity of Hurricane Katrina was at least in part a function of higher Caribbean water temperatures.

Let me be clear, though. I do not think that climate change necessarily means the end of human civilization, even if the climate is altered significantly. Indeed, this book does not recount in any detail the array of possible negative climate effects. Those possibilities are familiar enough, speculative in terms of detail, and readily available elsewhere.[2] A great deal depends on the rate of change and the human response to whatever changes occur.

The risks are nonetheless real. Humans, like most other species, have settled spaces and arrived at population levels based on climactic conditions in various locales. When those conditions change, the number of humans and the types and numbers of plants and animals that will thrive will also change, and only rarely will the number of species be increased, especially in the short term. It is that simple.

As adaptive as we are as a species, millions of humans may be unable to produce sufficient food or find sufficient water as and when climate alters significantly. At some point there will be enormous ecological disruption, and many species of animals and plants will be lost to extinction.

Some plant and animal species may be able to move northwards (or southwards in the Southern Hemisphere), but other species that already dwell in those spaces will be lost through an inability to adapt or compete with the new arrivals. In some cases we humans may be able to grow crops somewhat further to the north, or higher up on mountains, but for the most part such spaces suffer from very low soil quality because there has been no long, slow process of soil creation. The new agricultural spaces will be marginal, and more and larger spaces elsewhere are likely to be lost to agriculture through flooding or drought.

Most Canadians strongly agree that Canada should participate with other nations to lessen these and other impacts of climate change. Unfortunately, no one at this point knows if effective global climate change action is politically possible. China, India, Brazil, Russia, and other rising powers participate only nominally in the Kyoto Protocol, established in 1997 and ratified in 2002, and their demand for fossil fuels continues to rapidly rise. The United States signed the agreement but did not ratify it, and its government has so far done all it can to repudiate collective action. Nor was Australia fully on side; having negotiated plum terms (8 per cent *above* its 1990 rate of emissions by 2012), it declined to ratify until after the change of government in the November 2007 election.[3] Only Europe, Japan, and a few other nations are making concerted (and not always successful) efforts to reduce emissions. Even in Europe, the clear world leader on climate change, outcomes are uneven.

In many ways Canada's climate change story is perhaps the most curious of all. Canada agreed to a national target that was arrived at in what could only be described as an idiosyncratic manner. After great hemming and hawing and delay Canada eventually ratified the Kyoto agreement, but by the time it had done so the nation was a very long way – and perhaps now has become hopelessly – behind on the agreed targets. Canada then, generally for reasons unrelated to climate change, elected a government that never would have signed the agreement in the first place and in principle was ideologically wary of taking the action necessary to achieve compliance.

Opinion polls indicate that Canadians want a national effort on climate change. Many individual Canadians make an effort to make a small difference, especially with high energy prices also pushing them in the same direction. Although I won't go into the details – again on the grounds that they are altogether familiar and available elsewhere – the list of things that individuals can do is a long one.[4] This book focuses rather on the things that can and should be done *collectively*, and on how climate change really is an issue through which Canada not only can, but also inevitably will, define itself as a nation over the coming few years.

It could be said that Canada remains a nation unable to make up its mind, like the person in front of us in a lineup for ice cream or cold drinks on a hot day when we are in a hurry but badly want our treat. Canadians believe that *we should do something*, but do not fully know why the nation's output of greenhouse gases keeps rising. Canadians wonder what they can do as individuals, but yet the traffic flowing in and out of downtown Toronto or Calgary or Halifax just keeps growing. Nor do they tend to realize that the greatest increases in greenhouse gases are associated with the energy industry itself, and with our energy exports.

Canadians are also mostly too busy to be all that exercised about the level of inaction or the reasons for it. We feel sorry for the polar bears and vaguely sense that our nation's image is suffering, but really do not fully appreciate just how spectacularly our governments have dithered and operated at cross-purposes on this matter for a length of time that is now coming up on two decades.

Sometimes Canadians almost seem content in our near-total ambivalence and happily remain oblivious even to our recent national history on this issue. We fail to appreciate that Canada is in many ways crucial to the global outcome on this issue. We are also so used to thinking "I hope this winter is not too cold" that we fail to appreciate the many ways in which winter defines us. Even if that is not the case, the role that Canada plays in the global effort to avoid the worst effects of climate change will make it very clear where we stand and who we are as a nation.

Two

Canada, Oil, and the World's Stage

CANADA SOMETIMES SEEMS to have a national inferiority complex. Canadians lament being inconsequential and wish that they had the power to make the world a better place. Such is the fate of middle-sized nations.

But Canada, it turns out, has an opportunity to make a significant impact on the world at a critical moment in human history. Acting decisively with regard to global warming and the transition from fossil fuels could not be more urgent, and Canada is uniquely placed to make a difference. Canadians sense this, but somehow as a nation we have been all but immobilized regarding the issue for nearly two decades.

Canada is a big, cold, rich country that uses energy intensively, yet we have the collective capacity to turn our fossil fuel profligacy around in dramatic fashion, to demonstrate that this shift can be done even in a North American nation. If it can be done here, all will know, as they say, that it can be done anywhere. More than that, Canada's capacity to export energy could provide additional incentives to the largest energy user of all to reassume the global leadership role that it once held on environmental matters, on technology, and on marketplace innovation.

Canada, however, faces a dilemma – even an identity crisis (what else is new?). Unlike most nations in an age desperate for energy, Canada has a stunning array of options – and thereby a very real responsibility. If Canada cannot demonstrate the way to a viable post-oil energy future in timely fashion, then it is unlikely that any nation can. Yet Canada seems unable to act as a nation regarding energy matters, or climate change. The government signs a treaty to reduce greenhouse gas emissions and continuously acts, with encouragement and help from some provincial governments, in ways that promote a rapid escalation of emissions.

Canada's Energy Riches

Canada could be energy self-sufficient for centuries based solely on the supply of bitumen from the tar sands; it also has a large supply of coal. These energy sources are problematic in terms of climate change, but Canada has the technological capacity to learn how to use those resources responsibly in terms of greenhouse gas (GHG) emissions. If technological solutions are slow in coming, Canada has sufficient energy options, and Canadians have the sense of global responsibility necessary, to avoid further expanding the use of those resources until ways can be found to use them safely.

The government of Ontario has pledged to phase out coal use in the generation of electric power, and most of the increase in tar sands output is slated for export to the United States. In terms of Canada's own energy needs, Canada must develop technologies that radically reduce the GHGs associated with tar sands extraction and a proposed new generation of coal-fired power plants.

Above and beyond that, few nations with Canada's level of prosperity and economic diversification enjoy so many renewable and relatively benign energy options. If any nation can, Canada could come to function well without using any fossil energy. Canada could assume a global energy leadership role by emphasizing non-fossil options (including energy efficiency) in combination with low GHG fossil possibilities – and by considering as well the possibility of exporting fossil energy only to nations that adopt a responsible approach to energy use.[1]

Canada has an abundance of non-fossil energy options. Its combination of available fossil and non-fossil options is unique compared to other industrialized nations. Canada has sought a distinctive role in the world – and it would have exactly that if it were to play an important role in leading the way into humankind's inevitable post-fossil future. Doing so would make a statement all the more powerful because, given our fossil fuel abundance, we do not *need* to seek that leadership. The ethical obligation to do so, however, cannot be ignored. Canadians are worldly enough to know that no nation is an energy island, especially in an age of global economic integration and climate change.

The situation is as simple as this: global demand for fossil fuels is growing rapidly, and both the supply and the earth's capacity to cope with growing use are shrinking. The world has understood clearly since 1988 that climate change is a problem, and yet the demand for oil stood at 66.6 million barrels per day in 1990 and had risen to 83 million barrels per day by September 2007.[2] Canada's relative energy abundance and broad-based prosperity could provide the capability to resist the pressures that push against responsible long-term energy strategies. *Some* nations must resist the temptation to continuously grab for the short-term gains associated with growth in the production and consumption of fossil fuels, leaving future generations to fend for themselves on an overheated planet. Yet too few seem willing or able to do so.

While poor nations often cannot do so, Canada could resist the pressures to overexploit fossil resources in the short term. Given our wealth, our technological capabilities, our array of energy options, and our global orientation, Canada is as likely as any nation on the planet to maintain a fossil reserve that will endure into the distant future – a future wherein vast numbers of human lives and human progress will almost certainly depend on it. Whatever energy options are developed in the future, important economic activities will not be easily sustained without some reserve of liquid fossil fuels – air travel and petrochemical production, for example.

That reserve will also be essential to nations with fewer options and lesser capacities than Canada is fortunate enough to possess. Those oil-rich nations governed by dictatorships, or pressed by widespread poverty, are far less likely than Canada to act prudently and in the global interest. They typically do not even make choices that are in their *national* interests. Canada's democratic instincts, its wariness regarding hyper-individualism, and its foresight and prudence born of a combination of prosperity and a harsh climate could help to prevent a potentially catastrophic global future. A cursory glance at global events thus far in this new century indicates how dangerous an energy-hungry world might become.[3]

Canada could demonstrate that there can be life after fossil fuel dependency and at the same time keep an emergency reserve that could in the future help to ameliorate economic disruptions and maintain those activities for which post-oil alternatives will prove to be especially challenging. If a nation that does not need to make the transition puts itself on a path towards a post-oil world, other nations for which that path is even more necessary may well be motivated to accelerate their own transitions. What Canada, a large, cold country, can also demonstrate is that while moving towards a post-oil future may carry significant economic costs, it is not only possible but could also be reasonably comfortable.

Some of Canada's Post-Fossil Options

Beyond its formidable fossil energy reserves, Canada has an abundance of hydroelectric power, some still as yet untapped. Most regions of the country also have a great potential for wind energy – along the nation's extensive three-ocean coastline, including the Great Lakes, as well as in the Prairies and vast Northern expanses. To appreciate the scale of Canada's wind potential, consider this: Denmark envisions producing a significant proportion of the electrical energy that it needs from the wind on its coasts. All of Denmark, along with the wind-rich regions of Spain and Germany (Europe's other big wind producers), could be misplaced in Canada and overlooked for years.

Canada has an astounding 243,792 kilometres of coastline. Using only 5 per cent of that coastline and placing only ten five-megawatt (MW) wind turbines per kilometre (or a larger number of smaller units) would result in a capacity of 600,000 MW of electricity. That is equivalent to more than 120 Pickering eight-unit nuclear stations. That is far more electricity than could possibly be used within Canada.

Needless to say, a megawatt of capacity from wind does not yield an even flow of electrical energy at that level, but neither do nuclear stations yield their listed capacity all that reliably. Moreover, if we use only a small fraction of coastline (obviously far less than 5 per cent), we could select both high-average wind speeds and a diversity of windy seasons and times so that overall output would be fairly steady, especially when blended with non-wind sources. More than that, many non-coastal locations in Canada are very, very windy; output could thus be widely distributed throughout much of the country.

The biggest issue regarding wind is moving the energy to locations where it is in highest demand. One option is the development of low-loss direct current transmission lines.[4] A second is in some cases to locate the wind turbines within a reasonable distance of urban centres (100 kilometres is not a problem), to locate the turbines near to existing transmission lines from energy-producing remote locations (such as James Bay or Labrador), or to move high-energy-demand industries (such as metal smelting) near to wind-producing locations. A third possibility, in the longer term, would be to use remote wind farms to produce hydrogen, which then travels via former natural gas pipelines to points of use.[5]

Resolving both climate change and North America's strategic energy concerns requires that Canadians think big and long-term and consider our comparative advantages. Canada has an uncommon amount of land per capita, and that extensive space is a key to the possible development of several forms of renewable energy. Much of Canada's underutilized land could, for example, be used to produce energy based on biomass – especially the cellulosic options that, unlike most grain-based fuel options, have the potential to make a significant net contribution to global carbon balancing.[6]

Canada also has several other key advantages that are not often fully appreciated. The country is technologically advanced and wealthy and, most important perhaps, has a long history of willingness to marshal public resources to develop new technologies. Historically, Canadian public resources have played a central role in the development of hydroelectric power and nuclear energy. Those sources of energy reduce dependence on fossil fuels and as such may well have a role to play in the transition, but it is important to keep a level playing field among non-fossil options.

It is reasonable to argue that nuclear energy in particular should no longer be subsidized, and most especially it should not be subsidized more heavily than are Canada's many less environmentally problematic non-fossil options. Subsidizing fossil energy today (with the price of oil at $90 per barrel or more) is economic madness, a pointless giveaway. Subsidies make more sense for research and for the early development of high-potential, low environmental damage options that need an early leg up until they can establish themselves competitively.

Indeed, the option that governments should most encourage is energy efficiency in all its forms – both technological and social. That is, high-efficiency motors, teleconferencing technologies to reduce the energy used for travel, light bulbs and computers and bike paths, locally sourced food, and reconfigured urban spaces. Canada is again well placed for effective action on this front because it is a highly urbanized country. City dwellers typically live more energy efficiently than do suburbanites or rural residents, and they have the potential for even greater savings.[7]

Energy efficiency is the preferred option both environmentally and economically. Once saved, whether through technological innovation or changed habits, the energy use that is avoided will never again be needed – all of the environmental impacts associated with producing that energy are eliminated and the work that the energy accomplished is still accomplished. Even an energy source as benign as wind raises issues of competing land use, and biomass fuels often require considerable energy inputs (and produce GHG outputs) for harvesting, transportation, and processing.

The environmental impacts associated with producing an energy-efficient light bulb are little different from an ordinary bulb; indeed, they are lower because the bulb lasts longer. The energy used when someone cycles to work most days has little if any environmental impact, and the benefit comes not just from the fuel saved; the walker's tires and car will last longer and so will the roads, thus reducing GHGs even further.

The NRTEE Report: Using Less Energy Is Really Possible

The details on all these supply and demand reduction options provide great hope. Canada could sharply reduce its fossil fuel dependence in a matter of decades. The 2006 report of the National Round Table on Environment and Economy (NRTEE) on long-term climate change and energy strategy in Canada provides a careful review of GHG reduction technologies and possibilities.[8] Some commentators make significant GHG reductions out to be an enormous economic challenge, but that is not necessarily the case. The things that must be done are for the most part things that prudent managers might choose to do given rising oil prices and declining supplies, *even if there were no risk of climate change whatever.*

The world's reserves of conventional fossil fuels (other than coal) may soon reach a point where demand significantly exceeds potential output. Even if Canada has vast energy reserves we need to use those reserves as efficiently and cautiously as possible. How large a reserve Canada has relative to its own needs is almost irrelevant. Canadians live within and depend upon a globally integrated economy. There is no turning back, at least not easily. If the economy of China or the United States or another nation was severely hurt by energy shortfalls, Canada would not be immune from the result – whether through spreading economic contraction or wars for oil.

The human race is entering an era of energy scarcity – we can face it collectively or we can use energy shortfalls as yet another reason for conflict. The challenge is in some ways unlike any that humankind has previously faced. There is a real possibility that with lower levels of energy use from all sources, and if population levels continue to grow, the human population cannot prosper as much as we now do. Findings such as those in the NRTEE report or in books like George Monbiot's *Heat: How to Stop the Planet from Burning* suggest that the wealthy nations could prosper on less energy, but that is far less obviously the case for poorer nations that might well need to use more energy, albeit not necessarily from fossil sources, in order to obtain a standard of living that approaches what ours is today.

At increasing or even existing rates of extraction it is possible, given slowing rates of discovery, that it will not be very long before oil is less widely available. It is impossible to say exactly when we will reach that point, but it would seem that it is not likely to be longer than a decade or two into the future. Severe economic disruptions associated with energy supply shortfalls are a real possibility. The only rational response is to do whatever is necessary to reduce the extent to which our economies depend on fossil energy. Especially in nations that are now energy intensive it is crucial to maximize the economic output associated with every input of energy, as if everyone's grandchildren's lives depended on it, because in all probability they do.

Canada has energy reserves, but is nonetheless morally obliged to behave prudently and to reduce domestic fossil energy demand to the extent that it can. This is not only morally sound, but also economically sound, because every barrel that we do not use we can either export to those with greater needs or hold in reserve to sell at a higher price at a later date. Unless and until we are fully confident regarding alternative sources, we should not waste a drop of a commodity on which humankind's future depends. Canadians in particular need to remember that every gain in energy efficiency creates resources that can be sold at any time either directly or embedded in products that we might produce when few others will still be able to produce them.

Canada's moral obligations regarding energy extend even further. As an exporter of a commodity crucial to future economic stability, we are obliged to induce responsible energy behaviour in those nations that we sell our energy to. We are also obliged to be sure that our energy production for our own use *or for export* does not contribute unduly to climate disruption or to a possible very dangerous future of energy scarcity. That danger can be radically ameliorated by delaying the shortfalls further into the future to allow more time for both economic adjustments and new discoveries and technologies. In the present scheme of things, among all nations only the holders of reserves can ensure that future.

In this context the sorts of initiatives that the NRTEE report proposes seem the bare minimum that Canada might undertake. The NRTEE evaluates thirty possible GHG reduction options and concludes

that "there can be a domestic solution to making significant GHG reductions by mid-century" and "increasing energy efficiency is key – by
doing so we could achieve approximately 40 per cent of our goal of a
60 per cent reduction in GHG emissions."[9]

Indeed, looking item by item through this study, we can see that
only about one-third of the proposed GHG savings result from new
(mostly renewable) sources of energy supply. There are several substantial items in the mix that are neither new non-carbon sources nor
energy efficiency gains. For example, 5 per cent of the total gain
comes from switching to lower-carbon fuels (using more propane and
compressed natural gas and less oil and coal), and nearly 12 per cent
of the gains come from the development and introduction of carbon
capture and storage (all within the energy industry in Alberta and Saskatchewan).

The other NRTEE recommendations for energy efficiency improvements for the most part do not alter Canadian lifestyles in any very
big ways. New houses and other buildings are built to be more energy-
efficient, appliances and lighting are replaced with more energy-efficient
versions when it comes time to replace them in any case, and a significant proportion of existing buildings are gradually retrofitted with
more insulation and tighter windows. Tax rebates are instituted and
building codes are altered to help make these things happen. There are
still cars, but they are mandated to gradually, on average, become
more fuel-efficient, and public transit is improved so that more people
use it. Industries continue to become more energy-efficient, as many of
them have been doing for decades.

None of these things is an imposition, nor are any of them a threat
to economic well-being. Indeed, there might be more jobs associated
with the changes given that replacing light fixtures and windows is
more labour-intensive than producing the energy that might have been
used had these things not been done. We can also be pretty darn sure
that all of this is doable without threatening prosperity because Europe
is *already* roughly twice as energy-efficient as Canada (per capita or per
dollar of GNP) and all around every bit as prosperous.

In the NRTEE study the one-third of GHG savings from new sources of energy comes from a mixture of wind, nuclear, hydro, solar, wave/tidal, geothermal, biomass, and cogeneration (the simultaneous production of heat and electricity; it is also actually more an efficiency gain than a new source of supply). Much more might be done in terms of new sources of supply. For example, on wind energy the report assumes that "just over four times" the capacity already committed now would be added by 2050.

Thus this is really a fairly conservative estimate of what is possible in this time frame, and rightly so – any long-term study of this sort should be constructed using conservative assumptions. But we also need to think in terms of bold visions. The Green Party of Canada advocates a much more radical path regarding climate change, one that sees an 80 per cent reduction in GHG emissions by 2040, with a strong commitment to alternative energy sources like geothermal, cellulosic ethanol, and ocean-based technologies like wave power.[10] Some of these involve decidedly un-conservative assumptions about what is possible, but such visions are necessary to motivate a national will to continuously move forward.

Canada could be a world leader in the production of wind energy as well as a world leader in production of oil from the tar sands. Canada could in the shorter term delay the rapid expansion of oil production from the tar sands beyond the projects already committed – delay it until all future production could be achieved without GHG emissions greater than those associated with the production and use of conventional oil and gas (already a significant amount, of course). Why not require that extractors and refiners adopt sequestration or use non-fossil options to heat the bitumen?

The NRTEE report sees a 60 per cent reduction in GHG emissions by 2050, and more than that could clearly be done – and none of it would significantly harm the Canadian economy – if seen through continuously over that long time frame. Limiting tar sands production could slow economic growth in the near or even the medium term, but by 2050 output could be as large as it otherwise would have been given that limits like water availability will most likely cap total tar sands output in any case.

Tar sands projects that extend over nearly a decade have already been approved – which means that a moratorium on new approvals pending the availability of low-carbon techniques would result in only minimal delays. Even if significant investments in research and development on new extraction techniques and new technologies are imposed on the industry, at current prices there is a great deal of money to be made; and not only are oil prices virtually certain to increase further over a long time frame, but many of the innovative techniques that are developed could be adapted for other industries or used in other nations (Venezuela, for example, also has vast oil sands reserves).

The reason that a moratorium is urgent is clear. Once the capital investments have been made – resulting in massive GHG emissions – and the oil is flowing south to the United States, it would be very difficult to alter what has been put in place. Companies would argue, not unreasonably, that until they fully recover their very large investments, which might take thirty or forty years, they are not in a position to close down existing relatively new plants. More than that, the North American Free Trade Agreement (NAFTA) locks Canada into maintaining a level of energy exports once it is established.

Where will we be in terms of climate change or peak oil in the year 2050?[11] Do we want to preclude various options for future action in advance during the next decade? That is what we are in the process of doing by going slow on demand reduction and renewable energy, by not insisting on effective rules for tar sands development, and by allowing urban sprawl to continue unabated across Canada. We are doing that by locking Canada into additional energy exports to the United States without insisting on a commitment from our neighbour to participate in the necessary *global* effort on climate change. The next decade or two are crucial for both the world and Canada. Our own country, in particular, as we will see, has already delayed effective action on climate change for far too long.

Carbon Tax versus Cap-and-Trade

High energy prices in themselves will help to push Canadian citizens and Canadian businesses into doing the right thing – buying more fuel-efficient vehicles, altering the lighting fixtures in commercial locations, increasing investments in alternative (non-fossil) sources of energy. But rising prices alone will not tip the balance towards the decisions necessary for reducing GHG emissions. In some cases large corporations will have to be pushed into doing the right thing promptly, and this will have to be done through regulations that limit emissions or through market-based interventions that push the cost of emissions still higher.

One problem with depending on high oil prices to drive change is that we do not know if today's prices are a spike or a permanent change. The decisions necessary – to buy a hybrid automobile, replace all the windows in a house, live nearer to work, or buy a ground-source heat pump to replace an oil or gas furnace – are very big ones. People need to have a clear sense of where energy prices are going to be in five or ten years. That is even more the case regarding business decisions associated with a trucking fleet, a chain of retail outlets, an industrial operation, or a large public or private electrical utility.

In the case of the tar sands, coal-fired electricity, and other very large emitters, regulation may be the simplest option; but other policy tools could possibly be just as effective, perhaps even more so. The three main options pointed to in this regard are: regulation, carbon taxes, or cap-and-trade. The required regulation would set firm limits on emissions on a per facility basis; cap-and-trade sets a total limit for all large emitters and allows them to work out who does what and when; and carbon taxes would raise the price for all emitters to a level that achieves the same goal of reduction. Some critics argue that regulation is not the way to go, or that government micro-managing of producer and consumer behaviour through regulation is unnecessary and unnecessarily costly. Some say that the government has at its disposal other policy carrots and sticks that will work more effectively, and this is where carbon taxes and/or a cap-and-trade system come in. As Jeffrey Simpson, Mark Jaccard, and Nic Rivers argue in their book *Hot Air: Meeting Canada's Cli-*

mate Change Challenge, the best approach is a mix of all three policy options.[12] It is long past time to have put these options into practice.

Building codes also need to be toughened with regard to energy efficiency, and firm regulation is needed to gradually improve the average fuel efficiency of automobiles sold in Canada. Interestingly, a move by the United States to bring the fleet-average fuel efficiency of automobiles up from 35 to 45 miles per gallon would produce a savings of two million barrels of oil per day – as much oil as Canada consumes, and an amount equal to roughly two-thirds of Canada's oil exports to the United States (not that they would wish to diminish imports from their most politically secure supplier even if the demand was reduced by such an amount).

Carbon taxes apply a cost to all carbon emissions and place a corresponding tax (so much per tonne emitted) on the emitting fuel source at the point of production and/or use. Politically, advocating such a tax in the face of high gasoline prices is daunting indeed.[13] The Green Party of Canada is the only Canadian political party to include carbon taxes in its platform. It advocates a tax of $50 per tonne, which translates into $0.12 per litre of gasoline.[14] The Green Party platform offsets this tax by calling for corresponding reductions in income taxes, especially on low- and middle-income earners, and payroll taxes (employment insurance, Canada Pension).

Such steps are needed to offset the differential effect of a carbon tax on people who are less well off, who spend a higher proportion of their incomes on energy. There are other ways of approaching this problem, but it is disingenuous to pretend that there would not be a concrete social impact associated with policy options that significantly reduce GHG emissions. As far as the Green Party's choice goes, we need to remember that although payroll taxes are matched by individual taxpayers, they are capped and only paid on incomes up to a certain level. Thus lower- and middle-income earners pay a *higher* percentage of their incomes on these taxes than do upper-income earners. Reducing these taxes gives the less well off a bigger break. It should leave them no worse off even after they pay more for energy.

Carbon taxes can be offset in other ways. Certain incentives can reduce the need for energy, and in many cases these are already available – refunds on energy-efficient appliances, vehicles, insulation, and home renovations, for example. Another offset option would see only modest short-term carbon taxes (in the face of already rising oil prices), but those taxes would be designed to increase in the event that gasoline prices fell below a certain level. In other words, the tax might exist, but rise if oil prices were to abruptly fall at some point in the future (as they did in 1985, mostly in response to curtailed demand in the face of high prices). A carbon tax could thus be designed to put a floor on future prices. People and firms would then know that they must find ways of using less energy, and they would be less likely to make decisions based on the hope that prices are high only temporarily.

Even modest carbon taxes would be a very large revenue generator because they could be applied as well to energy exports. Whether the energy distributed is consumed here or elsewhere, carbon taxes discourage the production of carbon. They would be lower on natural gas than on coal, because natural gas produces less carbon. They would be non-existent on hydro power, wind energy, or nuclear power. To help keep transit fares low they could be waived for public transit companies and authorities. They would clearly help to change consumer and business behaviour. If they were used to replace other taxes, those removals would also change behaviours (lowering payroll taxes would encourage employment, for example).

Cap-and-trade is sometimes seen as an alternative to carbon taxes. A cap-and-trade system would also raise the price of emitting carbon by setting a maximum allowable total output and dividing the right to emit among emitters. Some proposed cap-and-trade systems would create a windfall by giving away some or all permits initially. Other systems hold an auction to sell some or all of the initial permits – which is the preferred option because it avoids giving a windfall to polluters and generates revenues that can then be used to offset inequities and encourage GHG reductions in other ways. In all designs, emitters can sell the permits they hold if and when they reduce their emissions; and the selling of permits generates revenue to offset or even cover the cost of reductions.

Setting regulations requires that governments either make decisions about technological options within particular industries or make distinctions about what might be possible in old as against new facilities. Allowing older plants to remain sheltered from the most stringent regulations because they cannot be easily adapted can lead to corporate decisions to leave them in operation. Carbon taxes and cap-and-trade systems leave the detailed decisions within industry and make them a part of everyday business decisions. At the same time they provide a financial disincentive to emissions and lead to revenue that can support governmental initiatives for individual Canadians and small businesses.

Carbon taxes can provide an incentive to action to all emitters, including individuals and small businesses (and even to others outside Canada that import our energy). Cap-and-trade systems provide greater emissions certainty than do carbon taxes because governments set the total emissions that will be allowed for large users (whereas carbon taxes provide cost uncertainty and no additional certainty as to exactly what behaviours those costs will induce).

Thus in many ways the various policy options are complementary. For those in power, in times of high and rising gasoline prices carbon taxes would be enormously unpopular, and a cap-and-trade system would be easier to see through politically. Carbon taxes, though, apply more broadly and could be used to put a floor on energy prices to avoid the sort of regression to energy inefficiency symbolized by the rise of the SUV after 1985. Carbon taxes would also be easier to administer globally – a factor that could prove crucially important in the future.

Let us imagine, though it seems hard to do at the moment, that both North Americans and Europeans dramatically reduce fossil fuel use in the coming decades. What would be the result economically? The November 2007 report of the Intergovernmental Panel on Climate Change argues that the costs to the world economy would not be all that large. As the *Washington Post* reported, "The most stringent efforts to stabilize greenhouse gases would cost the world's economies 0.12 percent of their average growth to 2050, the report estimates."[15] But as the demand for fossil fuels was driven down in the wealthiest economies through improved energy efficiency and by finding substitutes, there would be another impact: energy prices could fall or at least be kept from rising as rapidly as they might otherwise do.

That is one of the reasons why a global system is essential. With no GHG limits on China, India, Brazil, Russia, and other rapidly developing economies, these nations could take advantage of lower energy prices that could result from fossil energy use reductions in Canada, the United States, and Europe – should those reductions actually happen. Lower oil prices in 1985 were the result of efficiency gains, new oil discoveries, and use curtailment in the early 1980s. A similar result in the years ahead could undermine whatever initial GHG reduction gains were made.[16]

A reduced demand could lead to falling global fossil fuel prices (or at least a slowing of increases). Developing nations might in such a circumstance end up using more than they might otherwise have done. A global carbon tax requirement and a global cap-and-trade system would help to prevent that self-defeating outcome. The beauty of a carbon tax specifically is that compared to other policy tools it can be more readily applied globally.[17]

Canada and Climate Change

Continued stalling in Canada on effective climate change action is not just dangerous but utterly unnecessary. Not only does Canada have diverse options for taking a global lead on climate change, but it also has exceptional motivation. Climate change is highly visible in our far North, and our Canadian national identity is deeply invested in our Northern reaches. The polar bear has become a global icon of the ecological tragedy embodied in climate change. The undermining of permafrost is almost certain to seriously harm both the environment and economic activity in the far North. Moreover, the melting of Arctic ice in Canada's North, in Greenland, and in the Antarctic may be one of the greatest threats that climate change carries for human well-being everywhere across the planet.

Some sceptics imagine that Canada will somehow benefit from

climate change and could do with a bit of warming. That line of argument is plainly false because our ecology and way of life are adapted to Canadian weather as it has been experienced for centuries. Marginal increases in crop yields will not offset significant ecological losses, but that is only part of the story. Canadians may not suffer as much from climate disruption as others elsewhere – and particularly the worst-off people around the globe. The most vulnerable regions are dry regions, especially in Africa and areas that depend on glacial melt for fresh water, as in the Himalayas (and India) and Latin America. Next are the low-lying regions, including the island nations that are likely to disappear altogether. Compared to these threats, Canada will escape relatively unscathed. Believing that we will benefit, however slightly, or at least not suffer too greatly, is just flat out the wrong way to think about climate change.

Why? It is painfully simple. We no longer live in a world of separate, disconnected nations – climate change will have an impact on all life on this planet, human or otherwise, directly or indirectly, and Canadians will understand this if we just pause to think about it. To begin with, Canada is a nation of many peoples who have family ties to every corner of the planet. Flooding in Bangladesh, drought in Northern Africa, and the loss of glacial sources of drinking water in many nations will bring pain and suffering to the close relatives of many Canadians. More than that, we are quintessentially a trading nation. Globalization has created one world – what hinders or helps one nation will hinder or help all – and troubles elsewhere, including economic difficulties, refugees, and wars induced by climate change, could affect many of Canada's partners, customers, and suppliers.

Of course, the direct effects of climate change will also be noticeable, and damaging, in Canada – from North to South. The melting of permafrost may turn that vast Northern landscape from a carbon sink to yet another source of atmospheric carbon and methane, accelerating the climate disruption process. More than that, the need for limits on fossil fuel use will pose a significant challenge because Canada is still a cold country. We need to heat our homes and workplaces not just for comfort, but for survival. The energy required for this purpose may decline somewhat, but that need will remain even when other parts of the world become uninhabitable owing to drought, rising sea levels, or a loss of glacier-fed fresh water. Canada, as a very large country, must also move a lot of bulk goods (grain and ore, for example) to port, and bulk imports from coastal ports and limits on energy use may hurt us disproportionately, making us less competitive internationally.

Still, while being a big, cold country influences energy demand, big and cold are less important factors than one might expect. Our economy is, on the whole, integrated North-South rather than East-West, and most of the distances from Canadian production to U.S. markets are short. More than that, most of our individual travel patterns take us only short distances from our homes. Thus, when it comes to reducing energy the shape and organization of our cities are far more important than is the scale of the country.

The greatest Canadian challenges in the area of climate change are clear: our overall wealth and our energy production, especially energy production from the tar sands – an increasing percentage of which is exported. Wealth has meant that we drive cars as a matter of course, often big cars. We live in houses, often big houses. We consume a lot of goods, many of which travel great distances from around the world – including an endless stream of manufactured goods from China and out-of-season fresh produce from California, Florida, and, increasingly, Latin America and even North Africa.

But fossil fuel production, even more than fossil fuel use, is increasingly the root cause of Canada's inability to meet the nation's energy obligations. Vast amounts of greenhouse gases are released when we extract and process the fossil fuels that we later consume and export. Those fuels then emit a further measure as they are burned in Canadian and U.S. cars and trucks and furnaces. The economic boom centred in Alberta is creating a pan-Canadian environmental, moral, legal, and constitutional dilemma.

Three

The Tar Sands Dilemma

Do Oil and Democracy Mix?

DESPITE THE SPREAD of both oil wealth and democracy in recent decades, the world has very few oil-rich democracies. Is this a coin-
cidence? Norway and Canada are democracies, and so are Indonesia and
Mexico, though less reliably so until relatively recently. Venezuela and
Russia are struggling democracies – democracies arguably undermined
by their oil wealth (though one could equally argue that were it not for
that revenue, they would be shakier still). But, those nations aside, oil
wealth and democracy seem to mix about as well as oil and water.

One reason for this is that in today's world having oil results in
highly concentrated wealth under the control of either governments
or economic oligarchs or both. The wealth arises without advancing
economic diversification or providing broad employment opportunities. Those who hold oil wealth do not have a need to share that
wealth, at least to the same extent that the industrial barons of an earlier age did once unionization took hold.

Effective modern democracies in many instances emerged out of
the Industrial Revolution and the associated employment opportunities (however limited they were in terms of income at the outset) and
urbanization that emerged with it. That employment was concentrated in large workplaces, and those employees soon learned to band
together to advance their interests both in the workplace and at the
ballot box. In the contemporary world oil produces industrial-scale

wealth without producing employment or the forms of social organization that bring people together.

The oil industry itself produces very few jobs, and a high proportion of those jobs, especially the high-paying technical and managerial ones, often do not go to locals. The capital delivered goes to those who already hold power or to a small number of people who soon will hold power through the expenditure of their new-found wealth.

More than that, and perhaps even a greater threat to the emergence of democracy, oil is now such a strategically important commodity that its presence attracts the intervention of foreign corporations and foreign powers. Those external forces work closely with, and add to the power of, local political and economic elites rather than helping to create broad-based domestic power. Saudi Arabia, for example, is a case in point here. So is Iran, where democratic government was overthrown in 1953 and has not re-emerged since.

The scale of wealth now associated with oil is so concentrated that the industry can dominate a province, a nation, or a region economically, and in some circumstances that wealth can alter the character of political discourse. In Third World nations the oil industry engages in bribery, but in other nations (including Canada and the United States) it finances think tanks, political interest groups, political parties, and political leaders. This is not necessarily different from the ways in which any wealthy industry might operate except that, given the price of oil today, the oil industry often has the capacity to overwhelm all other sources of funds.

Another characteristic of oil wealth can also undermine effective democracy. The wealth arrives very rapidly, and it is known from the outset that this wealth is impermanent. This condition makes it very difficult for any given government to do long-term planning. Some jurisdictions, such as Norway and Alberta, have sought to build public trust funds to partially insulate public budgets against future revenue downturns as oil supplies decline. On the whole, though, the temptation remains to govern in imprudent ways and to avoid dealing with inevitable future downturns through a strong commitment to economic diversification. Future citizens do not vote in today's elections.

In all of these fairly obvious ways oil wealth can distort both economic and political development in places where such development is not well established. But even in a diverse economy such as that in Canada, where democracy is as well established as anywhere on earth, oil wealth can clearly influence the political process. Sometimes it does so in quite subtle ways that are altogether unintentional and are mostly just a function of sheer size and the temptations that sudden wealth almost inevitably carries with it.

An Exquisite Dilemma: The Choice between Wealth and Respect

The most obvious and the easiest route to increased Canadian wealth and power in the twenty-first century is through the extraction, refining, and export of fossil energy from the tar sands. As those that favour such development are quick to point out, the wealth generated by that development spreads to machinery suppliers and pipeline-construction equipment-makers in Ontario and elsewhere and into businesses of all kinds throughout the country.

Given the likely future price of oil, the almost certain continuation of political instability in the Middle East (and in other oil-producing nations such as Nigeria and Venezuela), and the scale of the resource in the ground under Northern Alberta, massive investment could continue to flow into Canada, and especially into Alberta. Interprovincial equalization payments and federal income tax revenues for that oil wealth may help extend the benefits of tar sands development to all Canadians through expenditures on infrastructure and social development. The temptations associated with oil wealth are obvious, not just for Alberta, but for all of Canada.

Fossil fuel resources, especially the tar sands, could help to ensure relative prosperity in Canada for another century. Given oil scarcity, our political stability, and the scale of the tar sands resource, both $100 per barrel and an output of five million barrels per day are within reach for the second quarter of this century (especially if adequate processing-water can be secured and environmental impacts are given minimal consideration). Thereafter, unless the global economy collapses, the price of oil can only go higher.

The value of an output that large could be as much as $200 billion per year, about 14 per cent of Canada's present Gross Domestic Product (GDP). And that money would multiply as it spreads through the economy. The people who arrange the financing, run the machinery, design and engineer the processing plants, make the steel for the pipelines, manage the oil companies, build the plants that heat the bitumen in place, and install and maintain the pollution abatement equipment and their families – these people will all need to be fed, housed, entertained, educated, equipped, and insured.

Still, a looming question hangs over this rosy picture. The question is: Could Canada exploit the tar sands at anything near to that level *and* meet its Kyoto requirements, let alone whatever future commitments might be necessary to reduce global emissions? The answer is almost certainly no, not unless the way in which tar sands oil is extracted and processed can be radically changed.

A corollary question therefore is: What will be the effect of this oil wealth and the economic temptation inherent in it on Canadians and their sense of obligation to the international community and on the functioning of Canadian democracy? Will Canadians continue to adhere to their internationalist, peace, and decency-oriented outlook, or will they join the very short list of rogue states that imagine that for some reason they are exempt from the needs and desires of humankind as a whole as expressed through international organizations and global negotiations?

Tar Sands: Making the Really Hard Choices

There is no avoiding that Canada and Alberta have hard decisions to make. The United States consumes about 24 million barrels of oil per day, but produces only about 6 million, and that amount is falling. All the rest is imported from the fifteen nations that are its major suppliers. Canada, not Saudi Arabia, is the largest single foreign supplier of oil to the United States. Indeed, Saudi Arabia is the United States' third-ranked petroleum supplier, at 1,394,000 barrels per day, behind Canada at 2,460,000 and Mexico at 1,538,000.[1] Canada supplies 11 per cent of U.S. oil consumption (and 16 per cent of its natural gas consumption).

At the same time many onlookers believe that Saudi Arabia exerts disproportionate political influence on the United States because of its oil reserves, its reliable willingness to supply oil, its influence over other oil suppliers, and its influence in the Middle East. Canada may lack Saudi Arabia's influence in the Middle East, but it has all the other components of presumptive influence. Yet somehow Canada is taken for granted.

Canada seems less able to lever significant influence for all its oil wealth and its continuing willingness to supply the United States almost exclusively. Canada, for some reason, does not yet think like an oil-rich nation despite having perhaps the largest long-term fossil fuel reserves in the world.

Oil today is the ultimate seller's market. Selling oil, even at high prices, is a favour to the buyer more than to the seller, especially if the capacity to supply that oil is long term and not subject to political instability. At the very least, Canada should gain political influence for its capacity to supply energy to those eager or even desperate to buy it (and that is a very long list).

If Canada were to decide to really be a global leader on climate change it could marshal the influence that arises out of its capacity to export oil. It could insist that the nations to whom it sells at least match its record on greenhouse gas reductions. This would only be an effective action, of course, if Canada had a record that we would want other nations to emulate. Needless to say, Canada would first need to make its own significant reductions – beginning in the industry that extracts the fossil energy that we export.

With oil at $90 per barrel, or even at $60 or $70, there is no reason that Canada's energy industry could not afford to alter the tar sands extraction process in ways that would radically reduce GHG emissions. Either the industry should be ordered to do so, or government should change the tax rules to generate the revenue to facilitate the changes itself. Government could, for example, fund the building of a pipeline to move carbon dioxide from Fort McMurray, Alberta, to a place where it could be permanently sequestered. But such an action should be unnecessary; regulation is the simplest approach and the most likely to succeed. Government could make drastic GHG reductions a condition of future production or export permits.

Canada in late 2007 was producing 3,135,000 barrels of oil per day – and the amount is growing even though Canada's supply and production of *conventional* oil are falling, as they are in the United States. Canada exports almost 80 per cent of total output to the United States and imports (in the eastern part of the country) enough to allow total consumption on the order of 2,300,000 barrels per day – slightly less than it exports to the United States. Total Canadian energy *production* has risen by 81 per cent since 1980, while *consumption* has risen by only 40 per cent. In the coming years production is touted to continue to rise rapidly while Canadian consumption outside of the energy industry itself will not necessarily rise at all.

Total oil production and exports are rising while conventional oil supplies are declining because of increased extraction of bitumen from the tar sands. This pattern could accelerate in the future. Massive investments are underway to increase tar sands extraction and Canadian petroleum exports. The scale of investment and rate of increase in output are either impressive or worrisome, depending on your point of view.

A 2005 study by the Canadian Energy Research Institute (CERI) estimated that tar sands investments between 2000 and 2020 would total $885 billion, with $634 billion of that taking place in Alberta and $102 billion in Ontario.[2] In 2004 some 41 per cent of Canadian oil production was in the form of unprocessed crude bitumen. By 2017 production of this material will have increased by threefold, and total petroleum output will also have risen even if conventional output continues to slowly decline. Total tar sands output stood at 1.1 million barrels per day in 2005, and CERI projected that amount to rise to 4.2 million barrels per day by 2020, well in excess of today's total petroleum production in Canada.

The tar sands development is obviously important, both to Canada's economy and to the overall security of energy supply in the United States. It is also central to Canada's prospects for meeting its climate change obligations in both the short and long terms. Why? Because petroleum products produced from the tar sands using present extraction and production methods result in twice or even three times the greenhouse gas emissions associated with oil from conventional sources.[3] This condition alone may all but determine the effectiveness of Canada's climate change performance. It is a critical factor: Alberta produces 73.2 tonnes of GHGs per capita, triple the amount per capita emitted in the United States and six times the amount emitted per capita in Quebec, where hydroelectric power is so important as an energy source.[4] As tar sands output sharply increases over the coming fifteen years, the level of Alberta's GHG emissions will almost certainly be even more out of line with the Canadian, North American, and global norm. If Alberta were a nation, it would be the largest per capita producer of greenhouse gases; it also would be the richest country in the world.[5]

Could tar sands development alone preclude Canada's meeting its climate change obligations? Put it this way: it will be very difficult to meet those national obligations if tar sands expansion is as great as anticipated and extraction technology does not change. Indeed, even the expansion so far carried out has contributed so greatly to putting Canada so far behind on its Kyoto obligations that there is little chance that these goals could have been met even if all other sectors had done their part.

While Canadians install new windows and insulation and buy new, more energy-efficient, front-loading washing machines and high-efficiency light bulbs and smaller cars and increasingly opt to take transit, emissions that offset all these individual efforts are scheduled to pour out of a series of new multi-billion-dollar tar sands plants. If Canadians want to reduce GHG emissions, they urgently need to understand this reality.

There is no avoiding the need to continue to do the many things that are recommended in terms of personal consumption if Canada is *ever* to get to the level of emissions agreed to in Kyoto. But we need also to either cap tar sands development or to find ways of producing fuel from that source without radically increasing emissions. It may come down to this: Canadians need to make a decision about how much oil and gas we can afford to export, both in thinking about our nation's long-term obligations to itself and to others and in meeting our global climate change obligations. To make such a decision we need first to appreciate the math. It is simple really.

If one million Canadians switched from automobiles to transit for most of their travel, the savings would be roughly one million tonnes of GHGs saved. If every Canadian household switched one incandescent bulb for a compact fluorescent bulb, the savings would be 400 thousand tonnes.[6] But at the same time the increase in GHG emissions from oil sands extraction and production (excluding the use of the end product in a vehicle or factory) by 2015 is anticipated to be 100 million tonnes. Clearly this would easily eclipse the amount that could be saved were every Canadian to switch to transit (tough to do in rural areas or small towns) and replace every light bulb, purchase a whole lot of energy-efficient refrigerators and washing machines, and install better windows.

Still, even if it were not for climate change all of the things that individual Canadians can do on a daily basis would be more than worth doing on the grounds of cost, health, declining global energy supplies, and air and water pollution; and the contribution of doing these things to reducing GHGs is far from incidental. The point is that as a nation Canada must make decisions about the extent of tar sands exploitation and the methods by which energy is extracted from that source.

At the same time we must face up to another large central source of GHG emissions from which almost all Canadians benefit. Canada now emits 758 million tonnes of GHGs from all sources. Our Kyoto target is 572 million tonnes (by 2012). The production and use of fossil fuels are responsible for 622 million tonnes (82 per cent) of the total emissions – the remainder comes primarily from land use changes and emissions of chemicals other than in energy production and use.[7] Clearly, the largest share of savings must come from this sector (along with savings from present and future tar sands extraction). This is where the importance of efficient household appliances and lighting comes in – they could help, along with wind energy and other supply sources, with the phase-out of some coal-fired power plants. But there is another option for some coal-fired power plants: carbon sequestration.

Carbon sequestration involves the capture of carbon from concentrated sources like the tar sands and coal-fired power plants and the subsequent pumping of it into storage. Especially promising is the possibility of permanent underground storage in aging gas or oil wells. Such locations are widely present in Alberta and Saskatchewan, but the redesign and retrofit to coal-fired plants (or the replacement of old plants with new) and the construction of pipelines and pumping facilities would be expensive, though perhaps not impossibly so.

Changes like these could add 20 per cent to the cost of electricity bills and are technically problematic in Ontario, where the nearest places for deep storage are probably somewhere in the United States, at a considerable distance. If the United States got serious about climate change, there might be intense competition for access to those storage areas, with U.S. utilities seeking ways to deal with capped GHG emissions.

The provinces of Quebec and Manitoba are more fortunate in that they have sufficient hydroelectric power that coal-fired stations are unlikely. Ontario's less expensive option might well be to phase out its coal-fired plants, as its government has promised to do in any case. The rapid expansion of the tar sands means that sequestration must be considered as one option for Alberta's many types of high-emission energy production and processing facilities.

But other carbon reduction possibilities exist vis-à-vis the tar sands. One possibility is to use geothermal energy – utilizing the heat in the hot rock deep beneath the earth's surface to warm the bitumen in the ground and facilitate pumping it to the surface. The warmed bitumen could also then be separated from the sand, using less energy from fossil sources. The heat is drawn from up to several miles below ground using water or other liquids sent through pipe loops, in much the same way as a ground-source heat pump might work, except on a larger scale and deeper underground, where temperatures are higher.

Another option is the use of nuclear energy as a heat source – a process that could be linked to the production of nuclear electricity or that could also utilize nuclear electricity. Using nuclear energy in a province where it has not been used previously may be a politically challenging decision, but it is not a decision that everyone concerned is prepared to exclude. Given the challenges associated with climate change, even some environmentalists would not exclude nuclear energy should neither sequestration nor geothermal energy prove to be viable options.

A question raised regarding the use of nuclear energy is whether it would be cost-competitive when all factors, including environmental risks and the insurance associated with the risk of plant malfunction, are included in the tar sands production process. While this position will not endear me to many of my fellow environmentalists, and while it would not be my first choice, I would not exclude the nuclear option out of hand – although I also feel compelled to point out that limiting annual output from the tar sands is also an option.

Nonetheless, several considerations suggest that geothermal energy might be the best option. For one thing it can be more easily produced in a distributed array rather than in one location – and less energy would then be lost in distribution. As well, given the considerable experience with deep-well drilling in Alberta, there should also be a capacity to achieve up-and-running status rapidly, especially as compared to nuclear power, which typically requires a decade of lead time.

This speed that the use of geothermal energy might allow is important in relation to the two vital objectives that are in sharp conflict here: 1) the pressure to produce energy from the tar sands as rapidly as possible, given the perceived urgent need on the part of the United States for more secure supplies of petroleum, and 2) the crucial role of the tar sands in relation to Canada's climate change obligations.

The tar sands plants that have already been approved well in advance of operation ensure, in the short run at least, that we are meeting the first of these objectives to the fullest extent possible. The only way of ensuring that the second objective will get anything like equal weight is to declare a moratorium on new approvals that do not address the issue of climate change in significant ways. The lead time associated with prior approvals may even be enough to ensure that there are no delays, because pressures to produce and export will ensure that the necessary technological developments will be accelerated sufficiently. If not, there might be a period of up to a few years in which no new tar sands plant comes on stream. Such an eventuality would hardly be the end of the world.

Environmental and other many organizations in Alberta and other parts of Canada have campaigned for a slower rate of tar sands development – though the news media have not widely publicized their efforts. Many of these groups support a moratorium on further development commitments. The Sierra Club, for example, is critical of tar sands development on a number of grounds, including the loss of boreal forest habitat – in order to strip-mine bitumen, large areas of land must be cleared of trees, undergrowth, and all plant and animal species.[8] Ecological restoration in the tar sands region, in the view of the Sierra Club, has been slow in coming.

Environmental organizations also raise the issue of water quality. The Sierra Club notes that if the tar sands tailings ponds (each an oily, sludgy mess) were combined they would comprise the third-largest dammed body of water on earth. Critics are concerned as well that water from the Athabasca River will not adequately meet the needs of tar sands expansion, let alone ecological and other human needs. For these reasons, and because of the contribution to climate change, the Sierra Club and other Canadian nationalist organizations based in Alberta, supported by other such organizations elsewhere in Canada, support a moratorium on expansion of tar sands projects beyond those already approved until Canada has a clearer overall policy on energy – especially regarding imports and exports. As Gordon Laxer, director of the Parkland Institute at the University of Alberta, put it: "We have been working on developing a comprehensive energy security strategy for Alberta and Canada. . . . Any such strategy must begin with a thorough analysis of the development of Alberta's tar sands."[9]

Bruce Campbell of the Ottawa-based Canadian Centre for Policy Alternatives put it more starkly: "We have less than a 10-year proven supply of both conventional oil as well as natural gas remaining yet most of the tar sands oil is earmarked for export to the U.S., and most of the natural gas from the Arctic – by way of the yet-to-be-built Mackenzie Valley pipeline – is also intended for the U.S. market or to fuel extraction of the tar sands crude."[10] *Fuelling Fortress America*, a report by these two organizations and the Polaris Institute, a nationalist-oriented policy think tank, calls for a five-year moratorium on new tar sands developments until a national energy strategy for Canada can be developed based on wide public participation.

The Polaris Institute hosts a website, *Tar Sands Watch*, which discusses tar sands expansion critically in terms of global warming, water depletion, Canadian energy security, Aboriginal rights, social damage, and concerns regarding U.S. militarism. In March 2007 it featured a radio interview with Grand Chief Herb Norwegian of the Dehcho First Nation, who joined in the call for a moratorium on tar sands development. "This so-called 'development' project is out of control. . . . It is like a cancerous tumour," Norwegian said. "The Mackenzie Gas Project is designed to feed that tumour."[11] Representatives of Alberta First Nations have also withdrawn from the multi-stakeholder Cumulative Environmental Management Association, a body created by the Alberta government in 1998 to deal with many of the environmental and other effects of tar sands expansion.

The Dehcho also oppose the Mackenzie Valley gas pipeline, which would feed tar sands expansion, until their land claims have been settled. More than that, the Dehcho are affected by alterations in the Athabasca River. As Tar Sands Watch puts it, "Elders and chiefs described how water levels have been fluctuating as much as 10 feet in some places along the mighty river and that fish and waterfowl are being negatively affected as well as wild game and the habitat they live on. The water is not fit to drink or swim in some places and fish have become soft and discoloured in others. The Dehcho rely on the water, fish, and game for food and trapping."[12]

Canadian nationalist advocacy organizations have long been

concerned about the undue proportion of Canada's energy resources that are being exported to the United States.[13] They argue that tar sands production will be committed for export while Canada continues to import half of its own requirements from outside North America. The fear is that Canadian energy imports will become unnecessarily exposed to the breakdown of supply from unstable locations, especially those in the Middle East. Most Canadian nationalists also want changes to the NAFTA agreement so that Canada would be allowed to divert its energy exports to meet needs in Eastern Canada should there be any threat of disruption of our oil imports.

Alberta's political parties are also involved in the intensifying discussion of the tar sands. Alberta's New Democratic Party (NDP) supports a moratorium on new tar sands developments until, as Environment Critic David Eggen put it, "A public commission has investigated all aspects and recommended a comprehensive strategy to deal with all effects, including greenhouse gas emissions." He also stated: "The Tory record on climate change is appalling. Under their watch, greenhouse gas emissions in Alberta have risen at a faster rate than any other province, by nearly 40% since 1990."[14]

The Alberta Liberal Party has also raised concerns regarding oil sands development, though it has not yet supported a development moratorium. Party leader Kevin Taft featured climate change front and centre in his 2007 alternative throne speech, stressing that on climate change actions Alberta was the worst-performing province in Canada and Canada was the worst-performing developed nation.

Taft also noted that Alberta was suffering the effects of climate change especially in the shrinking of the glaciers that supply fresh water to the province and in the increased number of pine beetles – insects that were once killed by Alberta's winters and are now threatening the province's forest industry. Taft called for hard caps on GHG emissions rather than intensity reductions, for carbon sequestration initiatives in the tar sands, and for increased use of transit and wind energy.[15]

Still, all of these moratorium advocates taken together have had only a limited influence on Alberta politics. They are easily outweighed by the combined forces of the oil industry, the Conservative party, and business interests from equipment suppliers to home builders and realtors, restaurant owners, and other retailers throughout the tar sands region. Most Alberta businesses assume that they benefit from tar sands expansion. In reality, though, many businesses are faced with labour shortages; manufacturing is hurt by a Canadian dollar pushed higher by capital imports and oil exports; and some Alberta municipalities, such as Fort McMurray, are experiencing growth rates that overwhelm municipal services.

All of these factors are important, but the bottom line when it comes to climate change is that using present methods of extraction, oil from the tar sands produces on average up to three times the GHGs per barrel that conventional oil produces.[16] *In situ* extraction uses roughly two times the natural gas that the surface-mining of bitumen does, and therefore, since the larger share of bitumen cannot be surface-mined because it is too deep, in the long run this GHG output will only rise unless extraction methods are radically changed.

That is the logic of a moratorium – to avoid sinking many billions of dollars into extraction and production facilities that produce amounts of GHGs that contribute to all but guaranteeing that Canada will violate not only Kyoto, but also the terms of all future global climate change agreements. In effect, improved U.S. geopolitical energy security is being purchased (and sold) at the price of climate security and Canada's international credibility.

This is not a trade-off that Canadians would be prepared to make if they fully understood it. Canadians are at heart more European than American in their approach to the world. It is also likely that the trade-off is unnecessary, given that the energy inputs to the tar sands need not be carbon-based; or even if they are, another option is carbon capture – wherein the carbon dioxide generated is used in existing conventional oil and gas operations to force additional increments of natural gas within older depleting wells to the surface.

The extraction and processing of bitumen involve three energy-intensive steps: heating the oil sands in place and/or mining them; separating the sand from the oil in a processing plant; and upgrading the bitumen prior to refining. The first two steps require low-grade heat, which can be produced without the massive use of natural gas, a premium fuel. As with coal-fired power plants, the carbon output within an upgrading refinery setting could be captured and sequestered, and the technologies to do so would be deployed in response to a carbon tax, an effective cap-and-trade system, or a regulatory regime.[17]

How Much Is Enough?

There are thus many arguments against unchecked tar sands exploitation. There are upper limits in terms of the water available to the production process, in the extent of the habitat of Northern Alberta that Canadians are prepared to transform into an open-pit mine, in the extent to which Albertans are prepared to put all of their province's economic eggs in one basket, or the extent to which the nation's exports are composed of one commodity shipped to one nation. How much habitat and water-quality damage should be borne by one region of one Canadian province? Climate change is but one concern, but it is an issue that is closely tied to the scale of tar sands development.

What are the appropriate limits to tar sands development? What is the best rate of extraction for Canada and the world, all things considered? How will this be decided? By whom will it be decided? The scale of tar sands exploitation should not be left solely to the marketplace, with the rate of extraction determined by the momentary availability of willing sellers and buyers. The issues involved are so large that the decision is unavoidably political in nature.

The decision, whatever it is, will define us as a nation for generations to come. The opportunities to widely discuss this issue are a measure of the quality of our democracy. Several important Canadian environmental organizations are involved in tar sands issues, including the Pembina Institute, World Wildlife Fund Canada, Greenpeace, the Sierra Club, and Environmental Defence. So too are other organizations, including the Council of Canadians. They have all produced detailed, carefully documented analyses of the issues, and outlined the decisions urgently needed.

To make these decisions we will need to weigh the global, national, provincial, and local environmental, economic, social, and political effects of tar sands development. The issue also involves longer-term considerations than markets are capable of handling. As the world is organized, responsibility for making these decisions and later, of course, remaking them rests with all Canadians and all Albertans and the democratic processes within which we all participate.

Something else needs to be said here. It is perhaps sometimes too easy to dismiss U.S. desires for importing oil from Canada as evidence of a voracious appetite for oil that supports extravagant lifestyle choices on the part of Americans, as if Canadians are somehow above such appetites. In truth there is little difference in per capita demand for energy in Canada and the United States. Suburbs on both sides of the border sprawl out from urban cores, and while big cars are somewhat more common south of the border than north, the differences are small – witness the proliferation of shopping malls and the traffic pouring out of urban areas on its way to cottage country on a Friday in the summer. We are obviously as prone to driving as Americans. We have signed Kyoto and they have not, but overall GHG performance has been no better here than there. Compared to Europe, Japan, and most other nations in the world both of our countries are energy gluttons.

Then too, many U.S. municipalities and states, including California and New York, have taken significant climate change initiatives. Perhaps the most stunning U.S. climate change policy failure has been the unwillingness of the U.S. federal government to force the auto industry to produce and market fuel-efficient cars and trucks. Nor has there been any U.S. federal leadership through energy tax initiatives or research funding on the scale that the problem demands. Moral leadership and significant and visible energy efficiency initiatives within government itself have been lacking.

Canada also has very good reasons to supply oil to the United States, China (another crucial trading partner), and other nations, besides the obvious boon that doing so provides to our economy. The reliable flow of energy from Canada to other nations contributes to their economic stability and in turn ours. More important, to the extent that Canada supplies base-level energy reliably, other nations are less tempted to rely on military might to gain future supplies or less likely to purchase those supplies from nations in which that money can leak into the hands of terrorists. In other words, to the extent that Canada can sell energy to the world, it contributes significantly to global economic and political stability. We cannot be naive about this. It is too easy to pretend that this is not the case.

At the same time, one nation cannot solve all such problems, and in a sense supplying energy to militarily powerful and aggressive nations is both enabling and soothing. The challenge for Canada is to use its capacity to supply energy as a lever to encourage international co-operation and even steps towards global governance. Especially given the urgent need for advances in global governance, Canada cannot allow supplying energy (or anything else) to other nations to undermine its capacity to participate in global initiatives (such as the Kyoto accord).

That said, there is more at stake here than at first meets the eye. This is not a decision in which all of the environmental weight is on one side and all of the economic weight on the other. All forms of energy production have environmental costs. All alternatives need to be weighed against each other. Few critics argue that all tar sands oil should be left in the ground. The questions are: What is the appropriate rate of extraction? What is the rate at which today's extraction should be increased, if at all? And what method of extraction and processing should be selected?

Most decidedly as well, contrary to appearances, all of the economic weight is not on the side of development. Important economic forces favour maximum development, but rapid development usually comes with unmentioned economic costs. Arguably the Canadian dollar has risen to the point at which our manufacturing industries and tourism have been badly hurt. That rise is in large part a function of the increase in resource prices, especially energy prices, and the rapid expansion of the energy sector and energy exports. On a local level within

Alberta, considerable wage inflation as well as inflation in housing prices and in the costs of all goods and services has occurred. These effects are transparent, but there are more subtly negative economic impacts associated with too great an emphasis on fossil energy production.

North American manufacturing is doubly threatened. Asian, Eastern European, and Latin American nations challenge us with manufacturing based on lower wages, a lower cost of living, minimal regulations, and minimal social benefits. North American manufacturing is also threatened by our own inability and unwillingness to adapt as rapidly as Europe and Japan have in the production of new energy-efficient technologies that create exports. These nations have adapted in part because they have not had our historic energy resource crutch.

The United States and Canada have for too long depended on fossil energy resources, and now the United States depends on imports from Canada and Mexico and on its military might to secure supplies elsewhere. North Americans have simply not had to think as long and as hard as Japan and Europe have about energy efficiency, or to act as comprehensively. We tax gasoline at far lower rates, and we have not made the effort that Europe has been making for decades. More recently Europe has had North Sea oil, but Norway has capped the rate of extraction to improve resource sustainability (also putting aside dollars for the future, to a greater extent than Alberta does).

As a result Europe has moved ahead and innovated in both energy efficiency and alternative energy options. The leading manufacturers of wind turbines are European, and great efforts are underway to develop wave energy. The most efficient household appliances are made in Sweden and Germany, and extensive public money is invested in public transportation infrastructure (though some of these systems are produced by the Canadian firm Bombardier).

Europe has even greater economic potential on this front, and by happenstance, nearly fifteen years ago, I came to appreciate that potential even more. At the time I was chairing the Mayor's Committee on Sustainable Development in Peterborough, my hometown. We were evaluating local businesses that were innovative in terms of environmental sustainability and presenting those that demonstrated excellent progress with local awards. One winner was a large commercial laundry that serviced a number of hospitals in our region. The firm had purchased new equipment that was valued at many millions of dollars and had cut its energy bill in half or better by, for example, recapturing all of the heat in the hot water being used – and very hot water was crucial for sterilizing patients' sheets, hospital garb, and surgical gowns.

That equipment was imported from Germany, where industrial wages are even higher than they are in North America, because that was the only place on the planet that produced it. Even at the high price paid for the equipment, the energy and dollar savings were dramatic, in part because the alternative was the purchase of expensive (and environmentally doubtful) sterilized and packaged throwaway gowns and garb. Germany had had limited energy options, and it was importing oil from the Middle East. Therefore the Germans opted to accept high energy prices sooner than we did, and as a result began to produce more energy-efficient capital equipment, which they were then able to export to the world.

Europe and Japan faced up to the challenge of energy constraints earlier than we did in North America. They did so out of necessity. Similarly, if Canada decided not to just fall back on the tar sands as an easy replacement for conventional oil and gas, the decision would not be based simply on being environmentally virtuous. It is also economically the better way to proceed. An economy that depends on easy, low-cost fossil energy will not produce goods of any interest to other nations, rich or poor. Chinese goods will be cheaper, and German or Swedish or Danish goods will be pre-adapted to the future. Unless a major effort is made very soon, North American goods will be neither.

We will never be able to compete with the Chinese, at least not until their wages rise to levels that approach ours, but Canada can find niches in technologies that enhance energy efficiency, produce clean energy, reduce GHG emissions, or sequester GHGs. Winter-climate wind turbines are but one real possibility with enormous market potential. So too are carbon sequestration or cellulosic ethanol, and both have great potential in Canada in the first instance. Other technologies with potential are ground-source heat pumps, light-rail transportation, low-energy computing devices, improved electric or hydrogen-fuelled buses, communications systems for door-to-door transit options, and all manner of more energy-efficient appliances.

Again, all of these options, other than carbon sequestration, would be worth pursuing even if climate change were not an issue. In any case, the global supplies of oil and natural gas are not adequate to reach rising demand, with or without the tar sands.

Options for Change

Much about the future of the climate is uncertain regardless of the actions that are or are not taken. We cannot calculate with any precision the pace of climate change under the various possible assumptions about future GHG emissions. There are just too many variables, especially in terms of feedback loops that accelerate or retard change.

We do not even know how great the effects will be from the amounts that have already been emitted. We certainly do not know what the effects will be in any particular place or at any particular time. Uncertainties are rife, and the natural human tendency is to think that it will all work out for the best, or at least that things will not be as bad as "some people" say they will be.

Reducing GHG emissions significantly may reach into the heart of our economy and alter some aspects of our way of life. Many things can be done relatively painlessly (change to better light bulbs and newer refrigerators), but many things will have significant costs. Many of us may need to find ways to drive less. Some industries may be severely hurt: electrical utilities, the coal industry, the trucking industry, or highway builders, for example. The risks are uncertain in some ways, and the known costs will impinge on different groups disproportionately. What is worse for the politics of the matter, the dangers are in the future and the costs of reducing risks are necessary *now*.

Given all of this I would stress three important points that we need to keep in mind. First, the process of change necessary to avoid the worst possible effects of climate change will take decades to see through – it cannot possibly be done in less time. Second, starting now and going slowly will prove to be far, far less costly than delaying and needing to make abrupt changes on an emergency basis somewhere down the line. Indeed, it simply may not be possible to make the necessary changes on an emergency basis. Third, most of what we need to do to lessen the risks associated with climate disruption, we would urgently need to do anyway even if climate were of no concern.

1. Making changes in the long haul. An enormous oil tanker cannot be stopped or turned around in less than a mile even if it is travelling slowly. Similarly, altering the production of greenhouse gases can be done, but not quickly (other than by hitting the metaphorical iceberg of economic collapse that might well be associated with *not* acting in a timely manner, and thus facing the worst effects of climate disruption). It involves altering the course of the economy, the ecology, and the climate of the whole planet.

To maintain a viable economy and comfortable lives without the massive use of fossil fuels, we need to make many, many changes. In North America we need, for example, to change the design of our automobiles, how often they are used, and how far they are typically driven. To do that relatively painlessly, we need to reconfigure our cities, an inevitably slow process. Given that we only replace about 2 per cent of the buildings and infrastructure per year, it will take decades to alter the shape of cities significantly.

Financial incentives for the use of transit (such as increased gas prices) will have little effect if people do not have options that are reasonably convenient – stores or jobs they could walk to in a short while or a viable transit system. Besides, changing our cities and our transportation systems is only a small part of what will be necessary to cut GHG emissions by half or more. People will not opt to heat and cool their homes with expensive ground-source heat pumps if they only recently purchased a new gas furnace or if they expect to be moving in a few years. Industries and cities will not choose to expend millions to alter energy use unless they are confident that energy prices will remain high or rise further. All of these changes take time. In some instances the process of change could be accelerated through tough regulations – but given the political challenges involved on that front, this option must be used selectively.

Even if effective incentives and disincentives are in place, the process will remain long term. Replacing recently purchased machinery, industrial facilities, appliances, and vehicles is in itself an energy-intensive action. We can perhaps speed replacement by a small amount, but most crucially we need to get things exactly right whenever the opportunities present themselves. Replacements must be made at the highest possible standard when long-range replacement cycles do come due – in homes, in cities, in industry, in commercial spaces, and in public institutions.

There are no crash-course emergency fixes. That is why Canada should not have wasted the two decades between 1988 and now, and especially the decade between signing Kyoto and now. That is why arguments that we need to wait for scientific certainty about exact effects or the effects in our particular location are utter nonsense. Nonetheless, prominent Canadian commentators still believe that this remains, even now, the most important thing to say regarding climate change.

Even if it is apparent that a wide range of possible climate outcomes exist, and even if we have only a vague sense about the timing of the various effects, we still need to start turning. We need to turn fast enough to avoid all of the worst-case scenarios even if they are fifty years out into the future. We need especially to put the brakes on in some directions, to not spend any more money on the worst-case options – like coal-fired power plants without carbon sequestration. When that money has been borrowed and spent, those plants (and all manner of other unwise investments) will be defended politically, and they will probably be used for most or all of their economic lives.

Further, it is always far easier and cheaper to build a house, a school, or a factory that is highly energy-efficient from scratch than it is to rebuild, to the same high standard, one that was not built to that standard in the first place. I am still a fan of architectural conservancy, but the price we pay now for inefficient buildings from the days of cheap energy is enormous. No matter how much money we spend we find it almost impossible to get them to standards that are fairly easily achieved when building from scratch.[18]

2. Starting now, going slowly, and avoiding an emergency. It is always cheaper to be prudent. Not just because the costs of climate disruption will far exceed the costs of avoiding it, but because doing things right the first time is far, far cheaper than fixing them after the fact. A highly energy-efficient house or appliance might cost 5 per cent more than a version of the same thing that was not energy-efficient. Fixing it halfway might cost two or three times that difference. Worse, having to scrap it carries the full cost at a time when we simply might not be able to meet that payment.

An imprudent corporation might be forced to scrap a ten- or fifteen-year-old production facility when Canadian society in general and government realize that this option is better both economically or politically than forcing people to sell off their cars. In a competitive world the cost could force the company to relocate production or cede market share to a competitor, or it could just sink the company.

An imprudent purchaser of a large exurban house a long way from anywhere – who later on tries to sell that house – might find that the house is worth far less than she or he originally paid for it. This is similar to the experience of the buyers and sellers of the fin-tailed auto behemoths of the 1970s and 1980s, when the cheapest second-hand cars were the largest cars. The huge vehicles that only a few years earlier were the envy of the neighbours eventually found their way into the hands of students and the poor because, for them, they were the only affordable cars (even if they could not afford to drive them much).

The most urgent policy step for governments at all levels to take is the establishment of tough energy efficiency requirements for all things new, including buildings, and especially the location of those buildings in relation to work and shopping. The smartest thing that consumers can do is to think about going beyond mandated minimums and paying a little extra for energy efficiency for the durable, energy-consuming items that they purchase. The smart thing for corporations to do – and to their credit many are doing very impressive things – is to design and build products now that meet the standards that will almost certainly exist down the road (and stop wasting money and time on fighting to force governments to let them do things as they have always done them). Most foolish in this regard was the North American auto industry, whose earlier political efforts to avoid fuel-efficiency standards increasingly now look to have been a death wish.

It takes time and money to learn how to do things right, but many businesses and manufacturers now understand that the way to future profits is through making those investments. Having the capacity to produce energy-efficient homes, trains, cars, and all manner of other goods – and to spend the money to learn how to produce them in an energy-efficient way – is a sound investment. General Electric is getting consistent 10 per cent annual profit growth from focusing on such initiatives. Toyota eclipsed its competition by means of an early entry into hybrid vehicles. Those corporations, and nations, that anticipate and do the development work avoid rising costs *and* end up selling their technological edge to the world.

3. We need to do it all anyway. I am not detailing what is known about the future of the global climate very much in this book, and that is because we do not know the details with any great certainty, and in any case they matter very little. The reason they matter very little is because *most of the things that need to be done to ameliorate climate change need to be done anyway*. Indeed, we have probably already used something very near to half of all the conventional oil and gas available to humankind for all of history. Climate change aside, we need to appreciate how long it will take to make a relatively smooth transition from oil dependency, and what are the consequences of failing to do so.

If the transition from oil and gas as our dominant energy sources to greater energy efficiency and post-carbon sources of energy is not smooth, some nations will be at least partially denied access to the remaining supplies of oil – by price alone if not by more overtly political means. Economies will be disrupted and economic dislocations will, in a global economy, tend to spread. More than that, if the result is economic turmoil, there is an even higher prospect of resource wars. Iraq is a clear indication of what is possible. So too are terrorist responses to, or provocations of, the occupation of oil-rich nations in the Middle East and elsewhere. All such dangers feed on each other and are likely to increase in the event of serious economic decline.

Clearly, it is time not so much to phase out the use of oil and gas as to start a long, slow reduction in the rate at which we use these precious substances. We can do this only by increasing energy efficiency, by avoiding unnecessary uses of energy, and by finding, developing, and utilizing a wide array of alternative energy options. Even if burning fossil fuels had no effect whatever on the climate, we would need to begin the transition in order to reduce the risks associated with anything but a long, slow, smooth shift in the energy basis of the human economy.

Climate Change and Peak Oil

Here we come to the essential difference between what needs to be done in the face of peak oil and what needs to be done regarding climate change. Given climate change we cannot simply replace oil and gas with coal (which is relatively more abundant, but considerably more carbon – and pollution – intensive). We need instead to concentrate on the cleaner alternatives: avoiding wasteful uses, improving energy efficiency, and finding and maximizing the adoption of alternative energy sources, especially renewable sources that are relatively benign environmentally.

That makes things harder, but not by all that much. Simply replacing oil and gas with coal is not in any case a reasonable way to proceed. Humankind in most of the developed world left coal behind as a home heating source and for many other uses for a reason. At present rates of energy consumption, if oil and gas were replaced by coal to any significant degree, the pollution levels, the land use impacts, and the occupational hazards associated with coal extraction would be intolerable. Again, even if climate change were not an issue, we might want to pursue the same course that the risks of climate change suggest: rapid improvements in energy efficiency and the expansion of renewable sources of energy (and, again, possibly nuclear power).[19]

A reasonable goal for Canada might be to gradually reduce total energy demand by half over a long period of time and to replace fossil fuels for half of the demand that remains. It is difficult to say how rapidly could this be done, but a number of studies suggest that forty years is an appropriate time frame: within that period we could maximize the smoothness and minimize the cost of the transition while respecting its overall urgency. Fifty-fifty by 2050 has a nice ring to it: that is, by 2050, through efficiency gains, we would use half of the energy we use today, and half of that would be met by sources other than fossil fuels.

That goal is possible: 2050 is a long time from now. To appreciate how long, think back to about 1965 and how different the world was then. Interestingly, the level of per capita energy use at that date was close to the 2050 target, but it was achieved in broad terms with unbelievably fuel-inefficient automobiles, more localized production of more commodities, far fewer trips by airplane, and the absence of computers and the Internet, which have considerable energy savings potential.

Significant reductions in energy use and fossil fuel use would doubtless affect the economy, but Canada would not necessarily be rendered poorer, especially in the long term. We would need to heat and cool buildings differently than we do now. We would need to make do with less transportation. We would need to transform our energy industries, and many others. All of these things are possible and do not necessarily involve great hardships if done gradually. The larger question is: Why do we seem to be in the process of blowing the options for smooth, gradual progress?

Increasing numbers of young urban Canadians already live lives without cars (other than using them occasionally, perhaps) and would not necessarily live less well in this future that might be. Many Canadians, of course, fear that future, and it is to that audience that Prime

Minister Stephen Harper appealed in early 2007 when he said: "I think the first realistic step in any . . . plan will be to try over the next few years to stabilize emissions and obviously over the longer term to reduce them. I don't think realistically we can tell Canadians: 'Stop driving your car; stop going to work; turn off your heat in winter.' These are not realistic solutions."[20] He did not say who it was that had urged Canadians to stop going to work or to turn off their heat in the winter. Neither the Green Party nor even the most over-the-top environmentalists have said any such thing.

Improved public transit and increased public funding to the point where fares can be reduced would lead more Canadians to seek new urban lifestyles – yet still might not match the subsidies provided in some jurisdictions in the United States, let alone Europe. Yet the prime minister adopted his ominous tone at the same time that he was placing climate change in a long time frame, and without proposing policies that might lead significant numbers of Canadians to make that shift. This political reluctance is not a matter of not wanting governments involved in markets – governments for decades have supported high-energy lifestyles through massive subsidies to energy industries, extravagant road-building, and in some jurisdictions tax patterns and zoning rules that virtually force suburban sprawl.[21]

Canadians actually do need to think in the long term even if the prime minister is reluctant to do so. We need to place the issue beyond the deadline prescribed in the Kyoto accord. For a century or more, most wealthy nations have grown economically faster than energy consumption has risen. That is, more and more goods and services are produced per barrel of oil or per unit of energy consumed from all sources.[22] That has been accomplished, in effect, without a systematic effort even when energy prices have fallen relative to the overall cost of living.[23] What is needed is a systematic effort and rising energy prices to accelerate that underlying trend.

Some forms of economic growth that might have occurred may be slowed or halted, but many other economic activities will be unaffected, and others will be pushed forward with policy initiatives that reduce the amount of energy needed to produce a dollar of GDP. Individuals will recalculate their everyday choices and, for example, opt to skip the week in Cancun and instead, at no greater expense, buy a new front-loading washing machine, join a nearby gym, or have several additional excellent meals in their favourite neighbourhood restaurant. In making these choices they not only forgo a very large energy hit, but their alternative spending could even result in actual reductions in total energy use.

Moreover, refitting public buildings or building new homes and offices to a higher energy standard would probably make the economy grow rather than shrink, and so too would producing more fuel-efficient cars, such as plug-in hybrid vehicles. Home energy retrofits produce more jobs per dollar expended than does supplying the energy. The primary cost to the economy would come in producing and consuming less energy, but that loss might well be restored because in Canada we would be producing and selling the energy we used (from wind power, small hydro, or nuclear) at a slightly higher average cost than we otherwise might have. The tar sands oil we did not consume would still be there to sell at an even higher price some years down the road.

As well, producing energy from the tar sands or sequestering the GHGs from a thermal power plant (and passing the 5–10 per cent cost increase through to energy consumers) would increase, not shrink GDP. We will use somewhat less of that energy if the price is higher, but that is all to the good in terms of climate change and peak oil, even if it is a little scary to petroleum producers – at least until they manage to think about it some more, and especially in the long term.

After the Tar Sands: What Else Is Necessary?

In the end, dealing with the tar sands (and coal-fired thermal plants) is the *sine qua non* of effective Canadian action on climate change. There is no avoiding that fact. The list of what else must be done is long, but everyone knows that and most Canadians are prepared to do their part as and when they can. With price and cost incentives, product availability, information, and encouragement, many, perhaps most, Canadians will make wiser choices when the time comes to trade in their cars, furnaces, or major appliances. In some cases those choices should be forced through effective regulation (minimum efficiency requirements or fleet average requirements for automobile manufacturers).

It is past time for Canadians to do their part, but it is also time to get beyond the implicit one-tonne-challenge illusion that Canadians can solve this problem through individual marketplace choices alone. Canada also needs to get past the tedious and relentless deregulatory mantra of recent decades. Regulations are especially important when it comes to large carbon emitters and goods that have a long life expectancy – especially vehicles, appliances, buildings, and cities. Minimum energy efficiency standards for buildings must be increased, because buyers have little influence over the standards to which houses are constructed. Green buildings are in vogue, but are as yet unavailable in most locations.

Home buyers have limited choices regarding where they reside, and many factors to consider (schools, price, resale value). It is difficult for energy issues to rise to the top of that list. The choice of where to live is easily as important in terms of energy use as is the energy efficiency rating of the house purchased. Especially within the middle ranges of possibility – that is, so long as a massively larger or massively more inefficient home is not involved – living closer to work and shopping is at least as important as the difference between one house and another. That too is in part a function of regulation – zoning to be exact. Mixed-use zoning and compact city design are essential to reducing energy consumption – and that is because, like buildings themselves, city form shifts slowly and we need to do things right nearly 100 per cent of the time when replacements occur.

Only a maximum of 2 per cent of housing stock is added in any year, and thus even doubling the efficiency of every unit built compared to the average saves only 1 per cent of total residential energy demand. Since it is far harder and far more expensive to improve the performance of existing homes, new home-efficiency standards are in order and so too are incentives to undertake expensive energy efficiency renovations. Regulations on renovations would also be useful.

There could be a requirement, for example, that an energy efficiency measurement of homes be done prior to resale and made available to all inquiring purchasers. The owners of homes that are out of line might then find that adding insulation and replacing leaky windows would boost their resale value.

Likewise, since the shape of cities also evolves very slowly, the zoning and building codes need to be correct whenever we build new homes and other buildings. Many people will choose to walk or cycle on nice days if the distances are short and the route is safe and interesting. Even if they drive, the distances traversed will be short if their homes are near to stores, schools, and employment. At even medium residential densities (a mix of single family and multi-unit dwellings), public transit can be economically viable, and, if it is sufficiently frequent, people will frequently opt to take it even if they own a car. This finding has been confirmed as a consistent pattern of voluntary behaviour in cities in North America, Europe, Asia, and Australia.[24]

Multi-unit dwellings are also typically more energy-efficient than are single-family dwellings. Accordingly such residential choices should be encouraged, especially when they are located near to urban cores where public transit is easily accessible and frequent and where walking to many destinations is commonplace. Incidentally, research has shown that urban dwellers have weight problems less frequently because walking is almost always a part of their everyday lives, even if they are transit users.[25] But at the same time there is an energy efficiency issue associated with many multi-unit dwellings: what might be called the landlord-tenant dilemma. If a landlord pays for heat and electricity, tenants have little incentive to conserve by turning down thermostats, closing windows in the winter, and turning off the lights. Studies have shown this difference to be significant. But, if the tenant pays, the landlord has no incentive to insulate or put in an energy-efficient furnace or new windows. The solution probably lies in rental property energy efficiency standards combined with tenant-paid heat and electricity (with corresponding rent reductions) or a mandated split of such costs so that both parties have an incentive to act appropriately.

Another realm where there is vast room for improvement is in the retail sector, especially perhaps in supermarkets and corner stores. As George Monbiot observes: "As you come through the door of a supermarket, a unit blasts you with hot air in the winter and cold air in the summer. . . . You must stand blinking for a moment as your eyes adjust to the lights. Then you walk past banks of fridges and freezers which have *no doors*. . . . But though you walk through valleys of ice, you remain warm. All day long, the freezers and heaters must fight each other."[26] The tale that Monbiot tells goes on at length and is every bit as true for Canada as for England. Some businesses in the retail sector have lowered the lighting levels and relocated the lighting fixtures (towards where the light is needed), but much more could be done and should be required.

Every sector from retail to manufacturing to extraction has room for improvement, and such changes are best achieved by the firms themselves, motivated by a judicious mix of regulations, tradeable carbon allocations, and carbon taxes that keep energy prices from falling when demand is reduced. The later item in the mix might be seen as *the lesson of 1985*, when improved energy efficiency and new sources of oil, both brought on by higher oil prices, resulted in falling oil prices. Major energy efficiency investments will not be made when businesses and consumers are uncertain about the continuance of high energy prices. We need to accept and learn to adapt to high energy prices if we are to cope effectively with either peak oil or climate change.

Another key is to find alternatives to coal-fired electricity as presently generated.[27] By far the largest GHG point sources in Canada are the Nanticoke coal-fired power plant in Ontario (17.6 million tonnes per year) and Transalta's Sundance coal-fired power plant in Alberta (16 million tonnes). These are followed by the Syncrude tar sands plant (10 million tonnes), EPCOR's largest coal-fired plant, also in Alberta (9 million tonnes), Suncor's tar sands plant (7.5 million tonnes), three other coal-fired stations in Ontario, Saskatchewan, and Nova Scotia, and the Dofasco Steel plant.[28]

Coal is a significant source of GHGs; a great reduction in Canada's emissions could be achieved if these facilities could either be phased out or the GHGs sequestered. Ontario has asserted an intention to phase out its coal plants in any case because they greatly contribute to air pollution, incidences of asthma, and acid precipitation. The province intends to replace them with improved energy efficiency, wind, small hydro, and an expansion of nuclear power. Again, Alberta, given the proximity of older oil wells and other locations where captured carbon dioxide might both serve a purpose and be trapped underground, has better sequestration prospects than Ontario has. Perhaps the most promising renewable alternatives to coal-fired electricity in Canada are wind energy and biofuels. Canada's middle and far north, as well as elsewhere in our vast topography, has vast wind-rich lands.

Biofuels are another relatively low-carbon energy option, and they include a wide range of possibilities beyond the widely known grain-based ethanol. While it may technically be the least effective option, politically grain-based alcohol is clearly the most appealing. It involves subsidies to farmers (always a political favourite) and, over and above that, helps to drive up grain prices and the value of agricultural land. If it is only added to gasoline usage (rather than replacing it), the auto companies, oil companies, or car owners will face no or only modest costs for vehicle or gas station adaptations. Technically, however, the net GHG savings in its increased use are very small, primarily because it takes a great deal of conventional fuel and petroleum-based agricultural chemicals to produce the alcohol.

There are many other biomass-based fuels. Good, old-fashioned wood for home heating does release GHGs, but will not involve a net addition if all the trees cut and burned are replaced with new trees. Pelletized wood makes for a fuel that can be automatically fed into a furnace while a homeowner is away and is also suitable for larger-scale uses. Wood and agricultural and forest wastes can also be converted to alcohols if technologies can be developed to break down the cellulose in those materials. Some pilot production facilities are under development, with aid from governments, but have not yet produced fuel on an industrial scale. Some critics remain sceptical about the prospects.[29]

Yet another possibility is the conversion of waste organic matter into methane, which is increasingly being done in Germany and elsewhere. This task has some special cold climate challenges, but feasibility studies are already underway in Canada. A significant advantage is that converting animal and agricultural wastes also produces fertilizer, without the necessity of using fossil fuels, and it reduces pollution and the contamination of groundwater in rural areas. It does not call for the use of additional land or drive up food costs, and it has the potential to produce an additional income stream for farmers. This mix of benefits suggests that public investments in this possible technological development are well worth making.

Many possibilities exist, but all of them require a considerable effort and probably also a commitment to determine and establish some limit to the development of easier, high carbon, environmentally problematic options such as the tar sands and coal. Few of these possibilities have been encouraged by government with much vigour or imagination.

Rather than steadily move ahead at the time when the problem of climate change became apparent, Canada has been dawdling for nearly twenty years – it has deferred, denied, and dabbled with the issue. Now we are left to wonder what a nice country like Canada is doing in a place like this: lagging behind even the United States in GHG performance despite having signed and ratified the Kyoto accord. Indeed, as we shall see, the climate change debate and climate change event hallmarks in Canada from the late 1980s until the present represent a sorry, though not uninteresting, tale.

Four

From Toronto to Kyoto on the Ambivalence Express

ANYONE REVIEWING climate change politics in Canada will be struck by the Liberal Party's history of ambivalence on the issue, especially when in office. They will also see the Conservative Party's history of denial, especially when *not* in office. When in office the Conservatives have been more accommodating, though that moderation on the issue was primarily during their former life as *Progressive* Conservatives. Prior to climate change being a household term, Brian Mulroney's government did what might reasonably have been expected from a government of the day regarding this issue.

Mulroney's willingness to act was perhaps in large measure because environmental issues were very much the focus of Canadians in the late 1980s and early 1990s during a deep and broad wave of environmental concern. It was also the case that since it was early days in the climate change game, nothing too politically difficult was yet expected of anyone. In 1988 the appropriate actions of the day were to fund research and to ponder low-cost actions that would challenge no entrenched political or economic interests.

Nonetheless, in hindsight a clear line is visible between the willing era of Progressive Conservative government and the ensuing climate change denials of the rising Reform-cum-Alliance-cum-Conservative Party of Canada. Although the possibility of human-induced climate change had been speculated about for a century or more, in the 1980s dealing with climate change had a "maybe someday" quality.[1] In the 1980s the idea of climate change was new and interesting but not very urgent, either as an environmental risk or in terms of the political costs associated with taking action. Issues like acid precipitation, toxic waste, the cutting of tropical rainforests, and the hole in the ozone layer all had far more resonance at the time. Much like environmentalism in its early days in the late 1960s, climate change in the late 1980s had an "oh-my-goodness-we-may-have-to-do-something-about-this" air about it.

Many environmentalists knew even at that early date that the costs of avoiding climate change were likely to be high. They recognized that GHG emissions were permeating the economy and that purely technological fixes would probably be insufficient. Whereas most forms of pollution required abatement expenditures by one or only a few major industrial polluters, climate change would require action by anyone and everyone. Environmentalists anticipated that while technical changes bringing about more fuel-efficient cars, buildings, and appliances would be important, behavioural changes would also be necessary. Nonetheless, in the 1980s and early 1990s neither political leaders nor the general public appreciated the scale of the societal challenge that would soon be widely seen as necessary.

Canada's national involvement with climate change came to the world stage when the nation hosted the 1988 Toronto Conference on the Changing Atmosphere. This event immediately followed the unprecedented 1987 global initiative on chlorofluorocarbons (CFCs) established in Montreal, an initiative that banned the production and use of CFCs in developed nations by 1995 and in all nations by 2010. That event, more than any other, signalled that environmental issues were increasingly moving from the local and national to the global level.[2]

The Montreal Protocol on Substances That Deplete the Ozone Layer, as it turned out, came to serve as a model for the scientific community for dealing with climate change. Scientists understood that to bring about concerted action on climate change, where the potential costs involved were far higher and political consensus much more difficult to achieve, they would need to reach a clear consensus. Anything less would open up an even greater chance of endless debate and ob-fuscation financed and encouraged by those economic interests that expected to bear some of the costs of change.

The 1988 Toronto climate change meeting, with delegates from forty-six countries in attendance, marked a major step towards creating a global scientific consensus and communicating that news to governments and the general public. What followed was a fairly direct path to the beginning of formal negotiations on a climate change pact and to placing the United Nations Framework Convention on Climate Change (FCCC) on the table at the same time as the Convention on Biological Diversity – at the 1992 Earth Summit in Rio de Janeiro.

All of these events took place during the Mulroney years (1984–93). Prime Minister Mulroney, addressing the Toronto Conference on the Changing Atmosphere, advocated an "international law of the atmosphere" accord, a proposal that reflected the language of the then-current international law of the sea treaty. Canada thus played a visible role in the early events that ultimately led to the Kyoto Protocol.

These events also coincided with what has been called the Second Wave of Environmentalism (roughly 1986–91), a period when public opinion and media attention focused unrelentingly on environmental matters – having been spurred by the issues of acid precipitation, the 1987 media sensation concerning the global journey of a barge filled with garbage from New York City, the 1989 Exxon Valdez oil spill, widespread public concern regarding the clear-cutting of tropical rainforests, and growing scientific evidence about human-induced climate change.[3] The Mulroney government responded to the rising public attention visible in polling results and media coverage with an emphasis on the issue of acid precipitation.

Once Ronald Reagan (who had consistently denied the existence of any and all environmental concerns) left office, Mulroney was able to negotiate and sign a treaty on acid precipitation with the United States. His fellow conservative George H.W. Bush was less resistant to action on environmental matters than was Mulroney's singing partner, Ronald Reagan, or than Bush's son would prove to be. But at a time when opinion polls showed environmental issues to be of greater concern to the public than jobs, inflation, or governmental deficits, even British prime minister Margaret Thatcher began to imagine for a short time that she was a green.

It is not known whether the differences amongst today's conservatives of various stripes, in terms of their depth of resistance to environmental concerns, is determined by the level of public interest or by deeply held ideological differences. That is, are Mulroney and the elder Bush, on the one hand, and Reagan, the younger Bush, and Stephen Harper (especially prior to being prime minister), on the other, different in their respective outlooks on environmental matters due to the varying moods of the day or because of differences between conservatives and neo-conservatives?

The very name *Progressive* Conservative was a Canadian curiosity born of a post-Depression era when anything called "Conservatism" was virtually unelectable anywhere on earth. About the time that Barry Goldwater and Ronald Reagan rode in out of the U.S. West with a full-throated conservatism, Canada too stirred yet again in its Western reaches. Economic libertarianism and deep social conservatism ascended as Alberta's gift to Canadian federal politics. These hard attitudes were served with a dollop of political populism and called Reform.

The actualization of neo-conservatism in Canada was most dramatically displayed following the election of Mike Harris in Ontario. Public budgets were dramatically slashed for services to the poor and environmental protection, both of which were seen as pernicious and misguided intrusions. Much was also made of privatization, including the privatization of drinking-water testing. Since the demise of the Harris days in the face of chemical spills and the Walkerton water tragedy, Ontario and much of Eastern Canada have been wary of neo-conservative assertiveness. Later, only evidence of corruption *and* spectacular dithering on the part of the federal Liberal government could induce even a halting public support for the new Conservative Party at the federal level.

In response, with advisors reminding him that Canadians are a profoundly moderate people, Harper adopted a breathtaking turnaround into seeming political blandness. That shift was certainly evident as well regarding environmental matters, but one is left wondering if the shift is more apparent than real. If Harper really is a reborn William Davis, what was all the fuss about that destroyed and buried the old Progressive Conservative Party? One is left to ask the obvious question: Why not just have continued the previous tradition? Conservatives prefer tradition, do they not?

Nonetheless, at the same time that the Mulroney government was playing an active role in the Rio Earth Summit, Harper was far from being bland in his politics. He exited the Progressive Conservative Party in response to what he saw as excessive governmental spending and a slowness on the part of Mulroney to revoke Liberal prime minister Pierre Trudeau's National Energy Program (NEP) – an attempt following the global oil price spike of 1979 to constrain the Canadian price of oil for long enough to allow the economy and Canadians to adapt. Indeed, Harper had previously exited the Liberals in response to the NEP initiative. Those who fought the NEP were opposed to *any* federal limitations on Alberta's oil industry.

Harper spoke at the 1987 founding convention of the Reform Party and played a significant role in its 1988 platform and in the creation of its slogan "The West Wants In!" The creation of the Reform Party was thus in large measure about the rise of the Canadian West and resentment over alleged attempts to limit that rise. Underlying this assertiveness, and an important dimension of it, was and is the unfettered expansion of the energy industry.

Interestingly, not every neo-conservative has consistently opposed all dimensions of environmental protection. Ronald Reagan did, but Barry Goldwater was decidedly ambivalent on the subject. Goldwater was a conservationist, but his economic libertarianism and anti-government inclinations tended to overwhelm this aspect of his belief system in practice. The Reform Party and the succeeding Canadian Alliance, however, at their roots leaned more towards Reagan's unabashedly anti-environmental views and less towards Goldwater's ambivalence. That approach is even truer of Harper than it is of Preston Manning, a Reform Party founder and leader, but both, like most Albertans, had severe reservations about Trudeau's National Energy Program.

With such post-National Energy Program pressures roaring to life out of the West, both the Progressive Conservatives and the Liberals were pressured to be wary of strong actions on the environment that might offend Western ranching, logging, mining, or, above all, oil industry interests. After Rio, environmental matters languished under both Mulroney and Liberal Jean Chrétien. Most notably, the Convention on Biological Diversity signed in Rio in 1992 with great flourish was not translated into a working Canadian Endangered Species Act until 2002.[4] As well, only very limited action on climate change paralleled this stunning decade-long hesitance to act.

Indeed, after Rio and throughout the 1990s the federal government was wary of taking a significant role in environmental matters despite the rise of involvement at the global level. Cutbacks and federal restraint with regard to environmental matters were the order of the day, part of a cycle dating back to 1968. In her book *Passing the Buck* Kathryn Harrison noted that in general the federal government was reluctant to be assertive vis-à-vis the provinces regarding environmental matters that were potentially within its jurisdiction.[5]

Specifically, under both Liberal and Progressive Conservative governments, the federal government avoided calling on the peace, order, and good government clause of the British North America Act as the basis for federal jurisdiction in environmental matters – an excess of caution and reluctance, some might say, since it is unlikely that anyone would have thought to list pollution or climate change in a set of rules dating to the nineteenth century. The inevitability of changing circumstances is the very reason why all constitutions contain such general clauses.

Mulroney's initial interest faded into ambivalence or indifference on climate change, and the rise of Reform Party resistance to Ottawa regarding the energy industry helped to keep it off the federal agenda for several years, basically until the lead-up to Kyoto in 1997. The early years of the Chrétien government from 1993 forward saw the federal reluctance to get involved in environmental matters reach new heights. As I wrote in a 2001 study regarding that era:

By the mid-1990s the green plan was effectively dead in a wave of federal cutbacks that hit environmental protection more than almost any other sector of government activity. Much of the 1990s retreat in federal funding for environmental protection (32 per cent between 1994–5 and 1997–8) was accompanied by rhetoric about eliminating duplication through federal-provincial 'harmonization.' Unspoken was the irony that many provincial governments, most dramatically Ontario, were cutting environmental activities at the same time (rather than picking up the slack).[6]

The Rio Conference, however, had resulted in a climate change agreement signed by 155 nations, including the United States. This broad-based agreement is now all but forgotten. The Kyoto agreement did not, as is sometimes now implied, come out of nowhere on a whim. Prime Minister Mulroney signed the United Nations Framework Convention on Climate Change at the Earth Summit in Rio de Janeiro in 1992. In that agreement Canada agreed to stabilize GHG emissions at 1990 levels by the year 2000. The 1992 agreement coming out of Rio was, as Deborah VanNijnatten and Douglas MacDonald put it:

> a "rather vaguely worded" document that acknowledged increasing concentrations of GHGs but required no action – countries were asked to voluntarily implement emissions reductions in order to return to their 1990 emission levels. Since Canada had already unilaterally made such a commitment, this international agreement was seen as relatively benign, and was quickly ratified by the Mulroney government.[7]

Canada thus remained consistent to its history of international cooperation regarding climate change, but also established a pattern of neglecting to see assertions and possible good intentions through to action. After 1992 the government, after consultation and internal discussion, developed a National Action Strategy on Global Warming, but little happened and emissions continued to rise.

While this pattern was soon to be repeated post-Kyoto, most Canadians have forgotten that our sign and ignore trick was a repeat performance. In 1998, after Kyoto was signed, the government created sixteen

Issue Tables involving input from 450 experts in what was called Canada's National Climate Change Process, which included a long series of public hearings. People of note from industry, government, environmental organizations, and universities were mobilized to participate. Extensive reports were produced, but again very little actually happened and emissions continued to rise.

This pattern of behaviour has revealed itself more than once when it comes to environmental issues. The public's concern regarding the environment is broad and real, but not deeply held and intensely pursued; the opposition to action on climate change is concentrated and intense. The solution thereby available to those in political office is to pay lip service – to offer the appearance of motion without the hard decisions and tough enforcement that would be necessary for successful action.

Political "leaders" manage to "show environmental concern" through meetings, discussion, protocols, treaties, and research, but never frustrate the strong economic interests that could suffer if strong actions were seen through into laws with teeth. In this same tradition is the passage of legislation with only minimal enforcement, voluntary actions, targets rather than strictly enforced regulations, and token prosecutions with minimal fines that are but a fraction of the cost of taking effective action.

If the public is insufficiently attentive because an environmental problem only impinges on small numbers of people in particular locations, or because its worst effects are both uncertain and set to occur at some unknown date in the future, this approach has worked very well. For a long succession of Canadian governments it has been almost standard operating procedure regarding climate change. Governments and entrenched economic interests can move very slowly on addressing the problem – or ignore it altogether – because of a seemingly short public attention span. To a certain point Canadians could be appeased by studies, conferences, new research initiatives, or seemingly tough (but generally not enforced) new laws. They could be reassured by global treaties involving very important people and very important nations. But very, very little actually happens – a lot of motion, but minimal results.

This is not always a deliberate strategy. It is sometimes more a matter of good intentions coming up against hard political realities – resistant corporations, provinces, or powerful nations that would prefer to avoid change. Following Rio, where Mulroney's political star was, to be charitable, fading, the Liberals under Chrétien campaigned on a platform of restoring the economy, dealing with government deficits, and protecting the environment. They published an elaborate policy Red Book emphasizing these features. On top of a general desire for a return to the political centre the conservative vote was then divided between the Progressive Conservatives (PCs) and the Reform Party, and the PC party had been all but destroyed in the election of 1993. The Liberal Party itself had its divisions, which also sometimes led to inaction. Sheila Copps, David Anderson, and Stéphane Dion as environment ministers and others, including Charles Caccia, were strong and effective advocates for action on climate change, but others in the cabinet and caucus were hesitant to take strong actions. Some just had other priorities; others just did not take the issue seriously. While the Red Book sought to one-up Mulroney on GHGs by advocating a 20 per cent reduction in carbon dioxide emissions from 1988 levels by the year 2005,[8] on election to a majority government the Liberals even failed to act in ways that achieved the more modest goals set out at Rio and signed by Mulroney.

VanNijnatten and MacDonald report that what ensued was detachment from the issue on the part of the prime minister and intense cabinet battles between Natural Resources Minister Anne McClellan (as an elected Liberal from Alberta, a rarity) and Environment Minister Copps. This interministerial struggle reflects a long history of how environmental issues have been treated in Canada at both the federal and provincial levels.

Most provinces as well as the federal government have seen long-standing tensions rooted in the cross-purposes between environmental and resource ministries – tensions that date to the origins of environmental ministries in the early 1970s. Historically the chief task of resources ministries was to advance the interests of and promote the expansion of resources extraction. The goal was to help to grow the segment of the economy with which that ministry was charged. The same would be true of an agriculture ministry or a ministry of industry or commerce.

Until environmental concerns developed and rose to a prominent place on the public agenda, few people objected to this approach. Government did what it could by way of infrastructure development, planning, research, and weather forecasting, for instance, to make it easier for businesses of all kinds and investors and employees to prosper. In the early days of the Ministry of the Environment in Ontario, the mining engineers and foresters of the Ministry of Natural Resources saw its employees as "hippies." There was a natural tension, albeit usually a mild one.

The tension arose because environment ministries are inherently different from most other ministries. Environment ministries sometime urge policies that impose limits on development or seek changes that will be expensive for industry. This is especially the case for primary producers (mining, forestry, fisheries, agriculture, and energy) that tend to be pollution- and energy-intensive.

Environment ministries inherently need to become involved in the affairs of every other ministry – a feature that tends to make many enemies within the bureaucracy. Environment ministries need to regulate agriculture in terms of pesticides, food quality, and soil and habitat protection. They need to constrain segments of the economy in terms of their habitual and costly behaviours. Effectively protecting the environment will almost always lead to stepping on toes. Like a finance ministry, an environment ministry needs to be at the centre of governance rather than be off on its own "doing its own thing."

The distinctiveness associated with environmental issues and ministries is true in spades regarding climate change. Every industry and economic activity, including those of government, is potentially significantly affected by climate change solutions – for example, the Department of National Defence or any provincial Ministry of Transportation or Municipal Affairs. This need to have a central position in the process of governance means that in a cabinet-style parliamentary government a prime minister or a premier may need to play an important role with regard to environmental issues. That was decidedly not the case in the early years of the Chrétien government.

In the Chrétien cabinet, McClellan, as the voice of Alberta and Natural Resources minister, was able to control the process of dealing with climate change. She could do this because Natural Resources Canada was in charge of implementation while Environment Canada had perhaps the larger role in international negotiations and some of the broader dimensions of climate change policy formation. Increasing Canada's energy production was another aspect of Natural Resources Canada's mandate, and that function would, as we shall see, eventually swamp government efforts regarding climate change. Moreover, the decisions regarding climate change initiatives following 1993 were made within a federal-provincial process wherein Copps and Environment Canada lost most arguments.

VanNijnatten and MacDonald noted, "The *National Action Program on Climate Change* (NAPCC), developed through federal-provincial negotiation under the aegis of the periodic Joint Meetings of Energy and Environment Ministers, was based almost completely on the instrument of voluntarism."[9] Not surprisingly, very little was accomplished. The much-touted voluntary measures were, in the case of climate change, a complete failure, with only a few exceptions.[10]

In its 1997 *Rio+5 Report*, the Sierra Club of Canada awarded the federal government a failing grade on climate change action. In the preceding year the Liberal Party had dropped "its commitment to reducing carbon dioxide emissions 20 per cent by 2005," and the Chrétien government had failed "to ensure equal access to tax advantages for the renewable energy sector as compared to the fossil fuel sector."[11] This was perhaps the lowest point in federal performance on climate change during the Chrétien years. Previous commitments had been all but abandoned in the lead-up to the scheduled meetings in Kyoto, Japan.

As the Sierra Club put it, Canada used the Kyoto lead-up process as an excuse to ignore then-current "voluntary" commitments under the Framework Convention on Climate Change. It was expected even before the fact that Kyoto would produce legally binding commitments to reduce GHGs in the post-2000 period. The Sierra Club concluded:

"Canada's position internationally on climate change has been based on excuses, self-serving rationalizations and a complete failure to take responsibility for its contribution to the problem."[12]

Clearly the Liberal government of the day believed either that effective action on climate change was a losing proposition in electoral terms or that it could finesse the issue with no one caring very much. Again, in this regard climate change was a classic environmental problem. Concern, such as it was at the time, was spread thin and lacked intensity, and the resistance to effective action was intense and willing to spend heavily on political action. The opposition was also highly concentrated in particular places, especially the West, where the Liberals were already vulnerable. And again, even more than usual the environmental risks were uncertain in terms of the particulars. The impacts were off in the future, but the remedy required that real and certain economic costs be imposed in the short term.

Then too, the early and mid-1990s were dominated by other concerns, especially in the period just following the Rio conference. The decade had begun with a deep recession, and the political agenda after Rio was dominated by the long pattern of high public deficits and rising federal debt. Some attention was paid to climate change even during the early days of the recession, but that attention involved little more than trying to understand a newly emerging problem and agreeing to very minor limits on emissions over an extended period of time. Indeed, during a recession reducing energy usage might have been comparatively easy. The Mulroney government, having been decisively voted out on other issues, was never tested in terms of actually achieving the reductions that it had agreed to in the FCCC.

In Chrétien's first term (1993–97) the focus was clearly elsewhere – climate change simply fell off the table. Consideration of all environmental issues waned, and government attention was transfixed on national unity, deficit fighting, and debt reduction. The low point on climate change (and the federal government's only F grade on climate change in the Sierra Club's annual report cards) was just prior to the convening of the Kyoto meetings. Mulroney had placed Canada at the centre of the earlier international process, such as it was, but with the waning of the second wave of environmental concern, as economic concerns regained centre stage, and before Kyoto was in their face, the Liberal government was unable to grant the matter any meaningful priority.

The rising indifference after Rio and the change of government in Canada were perhaps also in part due to Mulroney's enormous unpopularity at the time of his departure. His successor felt little obligation to abide by much of anything that the Conservative leader had done in his waning days.

The inattention also had to do with a cycle of alternating attention to economic and environmental issues that had operated since the 1970s – and the focus in the mid-1990s in Canada was on economic recovery from the free-trade adjustments that had hit Canada hard. The emerging economic recovery made balancing the budget possible for the first time in decades, and none too soon, almost everyone felt. Climate change, in contrast, still lacked a certain immediacy, and the Chrétien government acted accordingly.

Chrétien on the Road to Damascus

Thus for years Chrétien seemed, to say the least, ambivalent about climate change. It is unlikely that he thought about it a great deal. In his first term and beyond he was, above all, intent on Canadian unity, and with Paul Martin as finance minister he was determined to get Canada out of the spiral of deficit and debt. That debt had risen to the point where its trajectory endangered the Canadian economy and the future of Canada's social programs.

Chrétien also knew as well as any prime minister how to build and maintain an electoral base. Losing a few environmentalist voters to the NDP paled in importance to maintaining the government's credibility on fiscal integrity on Bay Street and Wall Street. He also had to take great care to avoid fuelling the fires of Western alienation, especially after having, only a few years before, come down hard on Alberta during the energy crisis days of the 1980s, in which he had himself played a central role. Moreover, he would be freer to operate in relation to Quebec if those Western fires could be abated. The National Energy

Program had been a significant political event for Chrétien. He was a key figure in the introduction of the program and in 1980 was decidedly unpopular in Alberta's oil country and elsewhere in the west. He invested considerable effort through the 1980s and early 1990s in undoing that intense dislike, a dislike often related to a general Western distrust of the East and especially Quebec.

In 1996, as prime minister, he went a long way towards finally overcoming some of that long-standing distrust. He travelled to Fort McMurray to provide some fanfare for new federal tax breaks for oil sands development. At the time Kyoto was still only a city somewhere in Japan – a conference was on the horizon but hardly at the top of most people's minds. Nonetheless, especially in retrospect, the tar sands tax breaks that Chrétien announced were staggeringly generous. They allowed oil companies to write off all capital investments in tar sands operations in the year that they were made. Companies with a tax liability could immediately eliminate the obligation. In effect, Canadian taxpayers would thereby pick up a significant proportion of the tab for all new tar sands operations.

Moreover, the investments that could be instantly written off could include cost overruns – a provision that was almost an invitation to extravagance and inflation. In support of the federal initiative the Alberta government allowed that royalties on tar sands extraction would only be 1 per cent unless and until the remaining capital costs were fully recovered. Those capital costs would not, of course, ever be recovered so long as continuous investments were made. This was a business strategy that made good sense in a highly profitable industry with too few other investment opportunities. Between 1995 and 2002 production from the tar sands rose by 74 per cent, but royalties paid to the Alberta government fell by 30 per cent.[13]

Alberta is now booming, but oil revenues to the Alberta government are far less than they might be. The oil is exported. The oil companies are predominantly foreign-owned. It all seems like an outrageous giveaway. The only factor that mitigates such a harsh conclusion is that when the arrangements were made the price of crude oil was far lower than it is today, and the prospects for massive profitability were far less obvious than they have turned out to be.

Oil prices spiked in 1979, but by 1988 (when the Intergovernmental Panel on Climate Change met in Toronto) they had fallen to about the same level as they were in 1978. Through the mid-1990s they were pretty much the same, and they were only around $20 (U.S) a barrel at the time Chrétien travelled to Fort McMurray. They only began to rise in 2000 and then exploded upwards when it became clear that the invasion and occupation of Iraq would be a very long way from the cakewalk that American neo-conservatives had imagined. But once in place, incentives to large corporations have a way of staying in place well beyond the point when an activity would be very prosperous even without the incentives. Chrétien never changed the incentive arrangement, nor did the government of Alberta change the extremely generous royalty regime until 2007, and then not by all that much.

Chrétien did come to think differently about climate change despite having probably, as Simpson, Jaccard, and Rivers contend, deliberately slowed climate change action between signing and ratifying Kyoto, leaving himself room to change his mind.[14] It was not scientific evidence, however, that persuaded him so much as a desire to keep Canada in line with other world nations – and perhaps to ensure that we were willing to go as far as the Americans were willing to go. As it turned out, of course, the United States never approved what U.S president Bill Clinton and his vice-president Al Gore negotiated in Kyoto.

As well, at the time, 2012 must have seemed a very long way off. Political leaders usually think in time frames closely tied to the next election. Chrétien at the time of Kyoto had just gone through an election and may have been uncertain about whether or not he would contest another. It would be interesting to know if the thought had crossed his mind that what he had promised to Alberta in 1996, perhaps as part of the lead-up to the 1997 election, might eventually render Canada's 1998 Kyoto commitments all but impossible.

Surely someone somewhere in the government must have had that thought, though it was probably still unclear at the time just how massive the uptake on tar sands investment opportunities would turn out to be. Political leaders and governments, like all of us, compartmentalize their various activities. In 1996 the government was dedicated to avoiding another referendum, having just narrowly avoided a loss in 1995.

The offer to the oil industry regarding tar sands investment would have been seen as doing something to offset the impression that the Chrétien government was only concerned about Quebec. Kyoto was primarily a matter for a different set of ministers and was seen as related to Canada's international image more than anything else. The linkages between the two initiatives might not have permeated the collective governmental consciousness all that deeply.

Chrétien himself was, in any case, not yet fully focused on his legacy. He wanted more than anything else to redeem himself for the near loss of Quebec to Canada and Canada to Quebec. Much of his life to that point had been dedicated to preserving the ties between the two. He was not about to walk away from government until that issue was further along the road to resolution, if not resolved. He may not have been paying enough attention to reflect for very long on the links between what seemed to be politically necessary tar sands tax breaks and an international agreement on climate change. Sign Kyoto, the Europeans want it, Clinton and Gore are going to agree too, we'll work out the details later.

The government's attitude at the time was revealed some years later in remarks by Eddie Goldenberg, perhaps Chrétien's closest advisor in the years leading up to and following Kyoto. It was the government's view that Canadians at the time only supported Kyoto "in the abstract" and that they were not then ready for the specific measures that would have been necessary to meet the Kyoto requirements. "Nor was the government itself even ready at the time with what had to be done," Goldenberg admitted in a 2007 speech. "The Kyoto targets were extremely ambitious and it was very possible that short term deadlines would, at the end of the day, have to be extended," he said, adding: "I believe that the signing of the Kyoto accord in the face of vigorous opposition served to galvanize public opinion to bring it to where it is today in Canada. In the long run, that will be far more important than whether we can meet the short term deadlines in the accord."[15]

On one level this is a reasonable statement, but on another it is an astonishing admission. Canada first became involved with climate change, recognizing the need for action, in 1988. Kyoto was agreed to ten years and *numerous* governmental study groups later. More than that, in October 1998 Robert Hornung and others working for the David Suzuki Foundation and Pembina Institute produced a detailed report on the actions that could be taken by Canada to achieve Kyoto compliance – a report that laid out a path that included fuel economy standards, a renewal energy portfolio standard, R-2000 efficiency standards for new housing, a modest gas tax increase (back before prices were as high as today), and a cap-and-trade system for industry.[16] Had some or all of that report been implemented, Canada would not be in the situation it is in today, a decade later.

In 2001 another Suzuki Foundation Report made it clear that increased GHG releases from the energy sector would make it very difficult to achieve Kyoto compliance.[17] Kyoto was ratified in 2002, with even more study and public involvement under the government's belt by that point, and around that time yet another detailed report was produced, this one by the Canadian Climate Action Network, laying out how Kyoto targets could still be met.[18] Clearly, there was and is opposition by powerful economic institutions to effective action, and clearly some in the public would flinch if their hydro bills were to rise or, in Alberta, if exploitation of the tar sands were slowed until technology could catch up.

These points of resistance could have been softened somewhere between 1992 and 2007, but they were not. No federal government has yet seriously attempted to soften public resistance – not Mulroney (even if he had done what might have been expected in very early days), or Chrétien, or Martin, or Harper. In the 1990s a concerted governmental and media effort convinced many Canadians of the necessity to drastically cut public spending, even of popular programs. Leadership can make a convincing case for many things if it chooses to do so.

Nor has any Canadian prime minister or leader of the opposition addressed the real source of Canada's failure to meet its climate change obligations – tar sands development without GHG amelioration and actual and proposed increases in energy exports. Most dramatically perhaps, the conservative opposition during the Chrétien years, especially the Reform Party and then the Alliance Party, flat out opposed any action by the government and denied for more than a decade, beyond all hope of credibility, the very existence of climate change.

Even as recently as 2006 Harper was still issuing semi-denials. At one point he referred to climate change in this way: "It is a complicated subject that is evolving. We have difficulties in predicting the weather in one week or even tomorrow. Imagine in a few decades."[19] That, of course, completely (and presumably deliberately) misses the point and the mounting evidence. No one claims that they are going to predict what the weather will be in any given week or year at some point in the distant future. Arctic pack ice and glaciers the world over are disappearing, and average temperatures are slowly but surely rising. Those conditions are well founded and not all that complicated.

The opposition that Chrétien was watching in the lead-up to Kyoto and in the long years of non-ratification that followed was the Bloc Québécois and the Reform/Alliance Party. NDP support was not growing significantly, the Progressive Conservatives were waning, and the Green Party was still under the radar. If there was a political threat, it would come from the far right and the West and move into Ontario, perhaps working in an anti-Liberal tacit ripsaw effect with the Bloc. It was therefore the interests of the oil industry that concerned Chrétien as much or more than any opposition that might be advanced by environmentalists or others concerned about global warming.

That consistent mood within the Liberal Party faded only briefly in Kyoto and in 2002 at the time of Kyoto ratification. David Anderson as minister of the environment and others tried to shift the hesitance to act, but rarely succeeded in breaking the spell.

Meet Kyoto: Not the Dog, the Treaty

Stéphane Dion's dog is named Kyoto, and Dion seems genuine in his commitment to action on climate change. However, the oil industry, the Alberta government, and the Reform-cum-Alliance-cum-Conservative Party still for the most part believe that the Kyoto accord is a dog that won't hunt. In their view Kyoto set Canada on an impossibly ambitious path, one that "does not allow for economic growth" – code for posing a challenge and an expense for those who benefit from rapid short-term expansion of tar sands extraction and processing, one driver of Canada's recent prosperity.

Maybe I am misreading him, but I do not see Dion as a hunter, and since Dion is not an American politician he does not have to pretend that he does that sort of thing.[20] As a Canadian politician he wisely based his campaign for the Liberal leadership on a determination to see an internationalist Canada willing to accept a firm national commitment to multilateralism and environmental protection. Multilateralism especially has been a Liberal policy hallmark for many decades.

The emphasis on multilateralism may in part explain the ambivalent behaviour of the Chrétien government. Throughout the mid-1990s the Liberal government was seemingly indifferent to domestic environmentalist pressures, but when acting on the world stage it was able to remember and act in ways that were consistent with Canada's historic internationalism. The competing domestic pressures in the 1990s may have favoured Canadian inaction on climate change, but the competing international pressures favoured action, especially during the late Clinton years, when the Americans appeared to be more or less on side with Europe regarding the issue.

As Elizabeth May has written regarding the lead-up to Kyoto:

Sheila Copps negotiated a mandate leading to Kyoto that established the importance of following the successful model of the Montreal Protocol. . . . To protect the ozone layer, the international community had agreed that the most important first step was for the industrialized countries, which had caused the problem in the first place and which had the resources to innovate and develop alternatives, to take on reduction targets, while leaving the developing countries to allow emissions . . . to rise in the short term. It had succeeded with the subsequent ozone protocols accelerating reductions in industrialized countries while bringing in the developing countries to cut back as well. The same approach was to be taken for the reduction of global greenhouse gases.[21]

Getting almost all of the nations of world to agree to anything is challenge enough. Getting virtually every nation to participate in an effort that involves enormous expenditures and could pose challenges to economic growth is near to impossible. Clearly a formulation that grants those least able to move ahead – and for whom *any* pressure on public budgets or *any* constraint on growth means that misery will increase and lives will be lost – is really the *only* possibility. This had been understood in international circles for many years, and both the Mulroney and Chrétien governments readily understood that this approach was necessary.

The time finally came, though, to settle on the extent to which the wealthier nations – those that had emitted virtually all of what had been emitted to that point – would cut. Who would cut what? Who would cut fastest? Europe wanted relatively more and had perhaps a better chance to achieve more on an overall basis as the inefficient heavy industries of Eastern Europe and Russia were wound down. This was especially true of Germany, which contained within it a radically restructuring East Germany. North America, Australia, and Japan (perhaps primarily because it was strongly influenced by the United States) wanted to go much slower, even though, other than Japan, they started from a much more energy-inefficient industrial and transportation base.

The Clinton administration faced particular problems: politically powerful industries such as autos and oil in politically crucial states, and a Senate that had signalled rejection of any agreement in Kyoto in advance of negotiations. Clinton himself was embroiled in all manner of political and personal difficulties and was focused on leaving a positive legacy nonetheless. He wanted an agreement if at all possible, but one that was not so stringent that there was no chance of getting the constitutionally necessary approval of the U.S. Senate.

May summarized events in and around the Kyoto meetings:

The provinces had been consulted and a target set of 3% below 1990 levels for our negotiating team in Kyoto. Chrétien was in Russia when he got a call from U.S. President Bill Clinton. Clinton reportedly asked

Chrétien to break a predicted impasse in Kyoto. The European nations wanted the protocol to mandate reductions on the order of 15%. Canada, the U.S. and Japan were only prepared to move much more slowly. Clinton asked Chrétien to offer deeper cuts in order to be able to achieve some success in Kyoto. Chrétien agreed, but allegedly obtained Clinton's support for some Canadian loopholes – credits for our forests (for the benefit of holding carbon out of the atmosphere) and for the export of greener technology. Chrétien's much reported "Beat the Americans" negotiating mandate to Canada's delegation was strategic collaboration with the U.S., not competition.[22]

Critics of the Kyoto Protocol have made much of Chrétien's alleged comment that Canada should "beat the Americans" by offering slightly greater cuts than they do. Clearly, if Canada offered deeper cuts than the United States, it would help Clinton in terms of his domestic political challenges in the Senate, and going to an amount higher than 3 per cent might help to keep the whole negotiating process from unravelling. In the heat of negotiations, adjustments must be made if any agreement is to be achieved.

Alternatively, Chrétien may have thought that if Clinton can go to N per cent with all of his troubles, then Canada could go to N+1 per cent. Besides, he might have thought, Clinton is going to help us out on getting sequestration in forests credited as carbon sinks and Canada has a *lot* of forests, more than any nation on the planet per capita. At the time he may not have appreciated just how far over the target Canada was about to go – far enough that we would have to plant trees in all of the Prairie wheat fields and everywhere else.

Either way all of this would be part of the give and take of any negotiating process. It would not be a casual whim on Chrétien's part, one launched in disregard of what Alberta premier Ralph Klein and others took to be quasi-sacred provincial rights and interests in the matter. In any case in the end Canada agreed to 6 per cent reductions and the United States to 7 per cent – we did not "beat them." And, to again quote May, "Canada was committed to 6% reductions, but as then Environment Minister Christine Stewart explained to angry provinces, the 6% target was really the same as the provincially agreed negotiation mandate of 3%. Once you counted in all the loopholes Canada had achieved with U.S. support, the amount of reduction required would be about 3% below 1990 levels."[23]

Whether or not the inclusion of credits for the purchase of carbon saving from poorer nations or the use of Canada's forests to sequester carbon are "loopholes," as May would have it, depends on the rules under which those credits are determined. The mere continued existence of forests should count for nothing, but afforestation – establishing forests on lands long lost to forests or never forested – will result in reductions in atmospheric carbon. Similarly, transferring energy-efficient technologies to poorer nations, where they can be installed at lower cost given lower land and wage costs, or where those proven technologies have already saturated our economy, is an altogether sensible way to proceed. The angels, as well as the devils, are in the details.

However complex discussions at the international level can become, and however difficult it is to attain agreements, in Canada federal-provincial agreements on anything are equally fraught, if not moreso. Perhaps that is why Canada produces so many skilled diplomats and imagines that international disputes should be resolved without resort to military forces. May notes that prior to Kyoto the 3 per cent figure that Chrétien was purported to have abandoned on a whim was arrived at following five years of federal-provincial negotiations. What other country on earth would have the patience to negotiate with its own provinces for five years prior to entering international negotiations, only to be attacked by those same provinces for having accepted a compromise that made the international agreement possible in aid of both Europe and the president of the United States?

Kyoto has been so maligned by Klein and George W. Bush and others for so long that we forget the extent to which it was truly a marvelous diplomatic achievement. It got developing nations that would otherwise have dismissed the project out of hand into the process. It has since been ratified by almost every nation on the planet save the giant holdout that has become a near pariah for having held out so long. An agreement was achieved that could lead to a process that could ultimately lead to the transformation, at great expense, of the core of all modern economies, and that agreement was fashioned within less than a decade of achieving the beginnings of scientific consensus as set out in Toronto in 1988.

The United States has argued that it should not be required to reduce its emissions so long as China and India do not. That is ethical madness. It is more surprising that China and India do not just say: *we* did not alter the atmosphere; whatever problem exists is not our doing. We demand that *you* fix it. Historically the GHG contributions of China and India and all the other non-Western nations on the planet amount to almost nothing (save perhaps for the effects of deforestation). Per capita China and India and the others *still* produce only a tiny fraction of what a Canadian or an American produces.

Kyoto is a global agreement, and everyone knows that such an agreement carries the implication that all nations will soon need to make changes from business as usual, but getting all of those nations to act requires that all of the rich nations first actually reduce their emissions. The United States has neither agreed to do so, nor done so. Canada has agreed to do so and failed miserably. It is hard to say which is worse, but in neither case is the failure the fault of a flawed agreement.

The Kyoto agreement predates the eloquent campaigns by Al Gore and others to spread the knowledge and to broaden public support for action in response to the scientific consensus. Much of the underpinnings of Kyoto were built up around the world one nation at a time prior to the full visibility of the extent of climate change that emerged through the 1990s as warmer average temperatures became more obvious. The agreement was achieved prior to wide knowledge of shrinking glaciers, prior to the highly anomalous northern winters of the early twenty-first century, and prior to the intensification of tropical storms and deepening drought in Australia and elsewhere.

Given that, Kyoto is an agreement that is in many ways unprecedented. It is an agreement that runs very deep, that at least starts a process that will be long-lasting. Moderating global warming does not involve or affect just a few polluters and economic sectors; it will reach to the core of every economy and at least to some extent change the behaviour of every firm and nearly every individual on the planet (save those who own little, travel little, and have little hope of ever doing so during their lifetimes).

The Kyoto agreement is a major step forward in internationalism and global governance. It sees almost all the nations of the world moving towards agreement on the management of the global economy. It could not come at a more important time in human history. Not only does the fact of global economic integration require that the nations of the world learn how to create meaningful global governance regarding tough issues, but in recent years the world has also moved radically away from rather than towards that possibility.

The world needs internationalism more than ever because a global economy requires minimum standards of workplace safety, pollution abatement, wages, and working conditions to avoid an economic race to the bottom.[24] A globally integrated economy means also that diseases that develop in one corner of the world move almost instantly to the whole planet, and therefore global health regulations are essential, not to mention rules that will restrain the evolving possibility of global terrorism made possible by the commonplace global movement of people.

Yet in the face of the clear need to create some workable system of global governance, the world has since the year 2000 moved significantly away from co-operation. The Geneva conventions, which have protected prisoners of war since World War II, have failed in Iraq and elsewhere. Nuclear proliferation continues, and nuclear disarmament and agreements on anti-ballistic missiles have been halted or reversed. The United States continues to reject the International Criminal Court. Only Kyoto stands as a recent move towards global co-operation. If the nations cannot agree and act co-operatively on a challenge as profound as climate change, it might be reasonable to conclude that there is little hope for the development of global governance. Without Kyoto, one would need to ask: Is there an international *system* in any meaningful sense of the word?

Five

Ratification and Beyond

Losing Face While Making Money

KYOTO AND ITS aftermath will say a great deal about where Canada fits into the world for the foreseeable future. Is Canada to be distinguished by a singular inability to live by its word? The United States and until recently Australia resisted effective climate change action, but they at least did not say that they would behave differently. Canada has continuously spoken in a clear and positive way about climate change action, and astonishingly, even with Stephen Harper, who began political life as a climate change denier, it still does. However, Canada appears unable to actually *act* beyond hosting conferences and signing international agreements.

Indeed, regarding climate change Canada sometimes seems not to be a nation at all. Canada appears, from the outside especially, to be a lot of economic and regional interests with no centre – incapable of seeing through on its desires or even its solemn word.

In the time from the first election of the Chrétien government until the lead-up to the Kyoto meetings, both the government and the nation were divided and ambivalent regarding what to do about climate change. At Kyoto Canada managed to play a credible and useful role, but thereafter the government lapsed back into squabbling with economic interests and provincial governments in the seemingly endless process. The nation practically staggered into ratification in 2002.

It has been said that national governments have been "hollowed out" by intense global economic competition and pressures from within from corporations (and in Canada's case, provinces) demanding greater autonomy from national rules. In Canada such pressures first arose most intensely from Quebec, but that province has now been surpassed by Alberta, whose government and bumper stickers once uttered the cry, "Let the eastern bastards freeze in the dark."

That was Alberta's response to the federal government's National Energy Program. As the province spoke, revealing its deep alienation from the rest of the nation, one could almost hear the echo of Bill Aberhart's voice travelling across the Prairies on a cold February night in the background.[1]

The internal wrangling had already produced five years of mostly closed-door discussions to establish a Canadian "national" position on climate change, and following the Kyoto conference the bickering continued. The federal government, having had the audacity to behave like a national government while bargaining on the global stage, came back home to internal heated struggles. Unending federal-provincial discussions: only in Canada, as they say in the tea commercials . . . pity. After Kyoto the oil industry resumed lobbying as only it can. The government of Alberta accused Ottawa of everything imaginable and some things not imaginable. The ever aggressive Mike Harris, still in power in oil-poor Ontario, sided with oil-rich Alberta more often than not, as did the other governments in provinces with fossil fuels (British Columbia, Saskatchewan, and Newfoundland).

The majority of Canadians wanted to see some action on climate change, but seemed not to be heard.[2] As with most federal-provincial wrangles, perhaps Canadians were too embarrassed or cross-pressured to speak out or even pay much attention to the argumentation. Nonetheless, through this period (1998–2002) Canadians increasingly favoured ratification of the Kyoto treaty. Dithering on the issue, however, became a national pastime. Little happened save for new variations on round tables and consulting vehicles and a few modest federal climate change funding initiatives.

Once out of the international spotlight the Chrétien government seemed to languish regarding climate change. From the ambivalence of the early days, before the Kyoto Conference, the government went to being publicly committed but unwilling or unable to push forward, especially once it became clear that the Clinton government had given up on ratification. Although the IPCC delegates who had assembled in Toronto in 1988, more than a decade earlier, had called for a 20 per cent reduction in worldwide GHG emissions by 2005, the Liberals remained irresolute on Kyoto, which demanded less than that of Canada in a longer time frame. At times prior to Kyoto Canada had seemed to stand with the isolated group of nations that were resisting action on emissions and against Europe and much of the rest of the world. The Clinton administration had not been in the vanguard on global warming, but was far from being the implacable object to change that the administration of George W. Bush was to become. Interestingly, Canada only moved towards ratification in 2002, once the Americans had become totally resistant and George W. Bush was emerging as hostile to *all* global governance initiatives – not just climate treaties.

In the four years that ratification stalled, Canada's GHG emissions continued to rise significantly. Indeed, it wasn't until 2004, twelve years after the first climate treaty was signed and six years after the Kyoto meetings, that the federal government launched CBC-TV comedian Rick Mercer in its One-Tonne Challenge ad campaign. That campaign was the first significant effort to convince the Canadian public that there were things that they should do.

The ad campaign was launched, however, two years after Kyoto ratification and eight years after the massive incentive program to the oil industry that had by then so increased Canada's GHG output that individual Canadians were all but powerless to bring the country back into Kyoto compliance. It was too little, too late, with a vengeance. It was also more than a little hypocritical coming from a government that had let large GHG emitters off the hook for more than a decade. By that point, though, those Canadians who wanted to see action on climate change were mostly grateful for anything they could get.

Along the way to that short-lived ad campaign there were more federal initiatives to offset the cost of energy efficiency improvements by homeowners and businesses, but no imposition of tough new rules regarding fuel efficiency in automobiles, or initiatives that reduced the cost of taking transit, or tough rules for utilities or other industries, and above all no GHG emissions rules regarding the rapid expansion of tar sands extraction. Alberta and the oil industry resisted all tough measures, and the Liberal government not only took no tough steps, but could also not even bring itself to ratify the treaty it had signed.

In the meantime, in 2001, within weeks of taking office George W. Bush announced that the United States would not ratify the Kyoto treaty, which he called "economically irresponsible." That assertion was made in the face of the conclusion of a massive array of notable U.S. economists that the treaty would do no irreparable harm to the U.S. economy.

Simultaneously in Canada the bickering intensified. From the lead-up to the signing of the Kyoto accord there was strong domestic resistance, but the pressure got more intense following Bush's ascendancy and explicit rejection of the accord. Ralph Klein of Alberta was especially fond of asserting that Canada could deal with climate change on its own. As he so delicately put it, "We don't need a bunch of international theorists to tell us how to do it."[3] Klein was working very hard to disconnect Canadians from their nation's long tradition of internationalism.

Sounding much like the arch-right of the U.S. South and Midwest, Klein fretted about "U.N bureaucrats telling Canadians how to manage our country." And, he argued in the lead-up to the ratification of the Kyoto accord, "Sacrificing jobs is entirely unnecessary, this is part of the problem with no consultation." His claim that job losses would be massive was presented without evidence. In 2002, regarding the absence of consultation, federal Environment Minister David Anderson noted that Ottawa by that point had been continuously talking with Alberta about climate change for five years. Anderson had in fact underestimated the length of the consultation process to that point by at least several years.

The days leading up to the ratification of Kyoto saw Klein reveal an utter absence of civility and national pride in a public display that rivalled his late-night drunken rant at an impoverished Albertan who had the temerity to be seen in his presence. In Russia, at a public event with the Moscow and international media present, Klein upstaged Prime Minister Chrétien by physically hogging the microphone. Klein was so combatively assertive that I recall thinking that if an American governor had behaved that way while the U.S. president was on the same stage, the U.S. Secret Service might have ended his performance rather abruptly.

As the Sierra Club put it in its 2002 Report Card on provincial governments, "The Klein government has sunk to new lows this year in its attempt to de-rail Canada's commitment to ratify the Kyoto Protocol. Ralph Klein's attempt to embarrass the Prime Minister in Moscow, waving about a letter the Prime Minister had not even seen and which did not say what he claimed it did, set a tone of disrespect breathtaking even from Mr. Klein."[4]

The notion that such behaviour was even possible – and that it did not result in national outrage – says something about Canada, a nation in which premiers become so used to "standing up against Ottawa" with moral indignation that one of their number is utterly unable to imagine how such behaviour looks on the international stage. But the scepticism regarding climate change was hardly limited to Albertans, oil executives, and arch-Conservatives – and it was a scepticism that ran quite deep in some circles until very recently.[5] Influential *Globe and Mail* columnist Margaret Wente, writing in 2005, wittily argued essentially that there was no hope of reducing greenhouse gas emissions. Voluntarism wouldn't work, nor would what she called bribes to buy "hybrid cars and solar panels." She wrapped up the range of possibilities with this suggestion: "The best way to get people to cut car emissions is to double gas prices, but that would be political suicide."[6] Shortly thereafter gas prices increased sharply on their own accord, without asking Wente's permission.

Before that event and after, Kyoto was, in Wente's view, "a folly from the beginning" and "dead as a doornail." In this foolhardy attempt to stop climate change, she explained, "We are spending billions on a scheme that amounts to nothing but feel-good PR and hot air, and has been rejected by almost everyone else. It is enough to make a girl crank up her hot water heater, turn on all the lights, and get in her SUV for a long drive to nowhere." If Wente believed that climate change did actually exist, she did not say at the time, though she has since allowed that it does. She, of course, really has (or had) an SUV and presumably enjoyed it. On one level what she argued was not unreasonable, especially if one assumes that "everyone" means "the United States." Wente never once complained, to my knowledge, about the billions of tax dollars that were forgone to accelerate tar sands development; she complained only about those tax dollars spent to discourage energy use. Spending billions of public dollars on schemes benefiting mostly foreign-owned corporations that export Canadian resources for which we are the stewards is presumably, for her, part of the normal scheme of things, even if it in effect subsidizes the forces that are altering the climate of the planet.

Thus Canada signed the Kyoto accord in early 1998, but somehow, incredibly, took until December 2002 to ratify the agreement. During the course of those four years there were many meetings, mostly behind closed doors, but no significant actions were taken. In the meantime Canada's emissions of GHGs rose rapidly as the economy grew and especially as tar sands operations expanded following the 1996 initiatives that provided the tax subsidy to investments that were guaranteed to produce massive new GHG emissions. In this period (1998–2002) the government's climate change actions (to use the word loosely) focused primarily on what was called Canada's Climate Change Process, a massive consultative undertaking.

If Kyoto was not a scheme that, in Wente's mind, almost everyone else had rejected, she might well have lashed out at the Liberals for acting at cross-purposes. She could have said that they were spending money to spin their wheels, that they were subsidizing the creation of more greenhouse gases in Alberta with one hand and talking everywhere else about how to reduce them. The net result was achieving nothing.

Along Comes Harper

In 1997, the year of the Kyoto conference, Stephen Harper decided not to run for re-election as a Reform Party MP from Calgary because Preston Manning seemed inclined to stay as leader. Harper was uncomfortable with Manning's willingness to take the party in a populist direction. He seemed to prefer that Reform remain a hard-core conservative organization even if that meant that it was destined to remain in a minority position.

From 1997 to 2002 Harper headed the National Citizens' Coalition (NCC), an organization working to advance a hard-core neo-conservative agenda in Canada. This he did throughout the time the Liberal government was hesitating and debating Kyoto, despite having signed it, and when the Reform party was firmly against ratification.

The NCC had been founded in 1967 by Colin M. Brown, an insurance company executive, to resist the establishment of a public health insurance system for all Canadians. The NCC also strongly resisted Government of Canada plans to admit Vietnamese refugees following the end of the Vietnam War. More recently the NCC opposed the existence of the Canada Wheat Board and railed against big government, unions, and the Canada Health Act. It advocated both tax reductions and a strong Canadian military. The organization played a significant role in the rise of Mike Harris as premier of Ontario.

Following his stint with the NCC Harper joined the Canadian Alliance (the successor to the Reform Party) and quickly became its leader. Soon into his tenure, in 2002, he penned a fund-raising letter on behalf of the party that focused on climate change and the Kyoto accord. In it he said: "We're gearing up for the biggest struggle our party has faced since you entrusted me with the leadership. I'm talking about the 'battle of Kyoto' – our campaign to block the job-killing, economy-destroying Kyoto Accord."[7] He certainly could not be accused of waffling on the issue, especially when he went on to say, "Kyoto is essentially a socialist scheme to suck money out of the wealth-producing nations."

The date is significant. The letter was written some four years after Prime Minister Chrétien signed the Kyoto accord on behalf of Canada.

Harper, proudly noting, "For a long time the Canadian Alliance stood virtually alone in opposing the Kyoto Accord," accused the Progressive Conservatives of speaking out of both sides of their mouths on the issue.

The fund-raising letter also includes an assertion that was to become central to his government's climate change policies when he became prime minister. He opined that one problem with the Kyoto accord was that, "It focuses on carbon dioxide, which is essential to life, rather than upon pollutants." For Harper carbon dioxide was not "a pollutant." One can only stand in awe at his concern for the respiratory needs of plants while ignoring the respiratory needs of the planet.

Despite the fervent opposition of the Canadian Alliance, the Kyoto accord was ratified by Parliament on December 17, 2002. The Liberal decision to proceed to ratification was difficult given the opposition from many provincial governments, the official opposition, the United States, and journalists like Wente and, from time to time, Rex Murphy. One important factor spurring ratification was that support from Canada and Russia was necessary to achieve the minimum support required by the treaty (without U.S. participation it was very difficult to attain support from nations responsible for the necessary proportion of emissions that the treaty called for).

Europe was pushing Canada to ratify and by that point Chrétien may well have begun to focus on his long-term legacy. It is possible that he resolved to decline Canadian involvement in the invasion of Iraq and to ratify Kyoto at roughly the same time. Perhaps he sensibly concluded that a measure of Canadian independence from a United States led by George W. Bush would serve his long-term reputation very well.

In spring 2003 the Progressive Conservatives selected Peter MacKay as their leader and the beginning of the end of the party was launched despite an agreement between Mackay and David Orchard, organic farmer and leader of the "red Tory" faction of the party. The red Tories were nationalists with a strong social and environmental conscience, a combination altogether out of tune with the political right in twenty-first-century North America (unless, of course, the nationalism was American nationalism).

Harper had already replaced the hapless Stockwell Day as leader of the Canadian Alliance and by the fall of 2003 the merger of the Progressive Conservatives and the Alliance was a done deal. The new Conservative Party of Canada was registered on December 8, 2003, with lawsuits pending from Orchard and other Progressive Conservative loyalists.

Possibly a fair number of the voters who supported the newly emerging Green Party came from genuinely progressive or centrist Conservatives suddenly left without a political home.[8] In a context of limited action on climate change from the Liberals and adamant opposition from the Conservatives, many Canadians were increasingly looking for a party that was firmly committed to effective environmental action and a moderate view on other issues. From that point forward the Green Party saw a slow but steady rise in popularity.

The Election of 2004 and the Conservative Victory of 2006

In August 2003, as Chrétien was preparing to step aside, presumably in favour of his arch-nemesis and sometime finance minister, the Liberal government announced additional post-Kyoto measures, including modest incentives on transportation choices, funding for initiatives by businesses, including tree plantations, improvements in the energy efficiency of federal government buildings, and partnerships with the provinces to encourage similar measures in provincial buildings.

What they did not announce were regulations regarding emissions in the energy sector either for the tar sands or coal-fired power plants, help for provinces that wanted to phase out coal, adjustments to the tax subsidies provided to the Alberta energy industry, or measures that would require higher fuel-efficiency standards for automobiles. Nor did they visibly pursue tough negotiations towards any of those ends. Spending initiatives were one thing; tough decisions with potential political fallout and economic consequences were another. We do not know if anyone in cabinet was pushing hard for any of the tougher measures.

In November Paul Martin was selected as Liberal leader and prime minister of Canada. After years of squabbling between the Martin and Chrétien camps, Martin took office only to face scandals of ferocious intensity arising out of events in the 1990s. The smooth Liberal machine sputtered and wheezed, the infighting continued, and corruption of bygone days in the form of ongoing revelations from the sponsorship scandal revealed itself relentlessly. A majority with 177 seats from the 2000 election begat a 135-seat minority in 2004, and in the election of January 2006 Liberal dominance fell away after eleven years.

Climate change did not have much effect in either of these election outcomes, nor for that matter did any substantive policy matter of the day. Canadians, especially in Quebec, were angry over tales of past Liberal corruption that centred on a challenge to Quebec nationalism. For all Canadians the Liberal image of inevitability and competent, smooth, and benign governance had dissolved. The relentless infighting continues to this day, to the detriment of effective action on climate change.

Voter anger and drift were the order of the day for several years, and Harper's Conservatives were the beneficiaries. Conservative vote gains were modest in 2004 and 2006, but Liberal losses – with a few soft votes also drifting to the Greens or the NDP – were enough to tip the balance of power into their hands. Voter distrust of the Conservatives on the issues of health care, the environment, including climate change, and relations with the United States remained for a majority of Canadians, but Harper and the Conservatives managed somehow to convey an aura of competence and seeming moderation, to at least some Canadians.

While it was in power the Martin government, like the Chrétien government, took additional public spending initiatives regarding climate change, but no regulatory initiatives, and even in the face of rising oil prices it made no adjustments to tax subsidies for the accelerating expansion of the tar sands. When he spoke in international settings Prime Minister Martin did, however, say the right things about Kyoto, and in March 2004 his government launched the One-

Tonne Challenge advertising initiative. As well, in early 2005, after being reduced to head of a minority government, Martin announced that later in the year Canada would host, in Montreal, the first Meeting of the Parties that had signed the Kyoto agreement.

The Meeting of the Parties took place, as it turned out, during a federal election campaign. Martin addressed the December meeting and stirred considerable controversy when he called out the United States by name for not participating in the Kyoto agreement. He was widely attacked for making a political opportunity out of a difficult diplomatic situation – although his remarks might better have been criticized for their naïveté about Canada's prospects for meeting its own Kyoto objectives. As Martin put it in Montreal:

> In October, I met with a group of Canadians concerned about climate change. They advocated short- and medium-term targets to guide efforts to reduce greenhouse gas emissions. . . .
>
> I'd heard these positions advocated before. But not from people like this. For these were the leaders of some of Canada's largest corporations, including those in the resource and energy sectors. They were encouraging Canada to adopt an aggressive plan to combat climate change. They had come to understand, they told me, that Canada's economic and environmental futures were entwined. And, more than that, that our nation had a responsibility to join those at the forefront of the fight against global warming.[9]

Cynics might say that Martin was disingenuous with regard to the views of these business leaders or at least willing to appear that way in public. He seemed to exclude the possibility that some of these leaders were telling him what he wanted to hear while they waited for the results of the next election. Alternatively, Martin might have been telling the conference delegates, and the Canadian public, what they wanted to hear while he knew full well that his government was systematically avoiding hard measures.

In the speech Martin also spoke of "working towards our Kyoto commitments" – as distinct from "meeting our Kyoto commitments."

He also touted "the greenest budget in our history and a comprehensive climate change agenda." That agenda did offer federal funding towards the development of alternative technologies, but did not require the adoption of the best technologies in industrial, commercial, or residential settings even at some date in the future.

Martin, perhaps pushed by Dion as environment minister, did seem to be working towards long-term GHG reductions, and in February 2005 he requested a series of studies and reports by the National Round Table on Environment and Economy. The 2004 federal budget included $200 million in funding for sustainable development technology research. The 2005 budget included $10 billion, an impressive amount, but it was to be spread over the years 2005–12. The money included funding for the EnerGuide for Houses program, climate technology research funding, and additional partnerships with provincial governments. The program mentioned a new regime for Large Final Emitters, but no regulations or enforcement mechanisms of any kind were forthcoming.

The attacks on climate change initiatives and the Kyoto agreement from the Conservatives continued, but they could not match the intensity of continuing attacks on Martin from within the Liberal Party. The level of animosity between those around Martin and those around Chrétien can only be fully appreciated by looking backwards from Chrétien's 2007 charge in his memoirs that Martin's hesitation to commit troops in timely fashion to Afghanistan caused losses for Canadian troops. Straight from the heart, straight from heart surgery, that charge was levelled long after Martin had left office. With friends like that it is little wonder that the Martin government was short-lived.

Again, it is also clear that the electoral outcomes in 2004 and 2006 had little to do with substantive issues, including climate change. Canadians wanted climate change action, but did not have that concern at the top of their minds on election days. Yet the outcome, strangely enough, would have a very clear effect in that regard. In April 2006 the newly elected minority Harper government eliminated or reduced fifteen federal government programs on climate change. First on the chopping block was the One-Tonne Challenge, a program that probably did miss a key point: that voluntary actions by individual Canadians would not likely be capable of offsetting a massive increase in the production of tar sands oil, an energy source that Elizabeth May called "the planet's most carbon intensive oil."

One of the other programs cut in April had a far greater potential for making a difference in the long run. The EnerGuide program provided financial incentives to Canadians for home retrofits – added insulation, new windows, and other energy efficiency renovations. These programs generally also had considerable local economic benefits because the renovations encouraged work for local contractors, carpenters, other tradesmen, and retailers. With only a small proportion of the cost of the renovations coming from tax savings or grants to homeowners, the government would probably have recovered much of the money in taxes paid by these economic beneficiaries of the program.

Around this time, as later revealed, civil servants were also asked to "remove references to Kyoto from the government's climate-change site in May and then to shut down the site, climatechange.gc.ca, entirely in June."[10] The government, it seemed, did not want to remind Canadians that it was obliged to take action on climate change under international law contained in a treaty to which Canada was a party.

Having apparently given up on flat-out climate change denial, the Harper government, like its predecessors, was now prepared to act in minor ways so long as the pressures to do something continued. Harper and his government were mostly concerned, however, about refocusing the attention of both Canadians and those in the international community who were aware of Canada's record of climate change inaction. The goal in terms of refocusing was to turn all eyes to the long-term future and away from the terms of present incumbents – a trick that should long since have become an obvious ploy.

The refocus campaign also included discussions of smog, as if it had something to do with climate. The Conservative government also spoke of "emissions intensity" rather than actual emissions, and about "made-in-Canada" strategies. But mostly Harper centred discussion on 2050 rather than on now or anytime soon. This ploy also included conveniently forgetting that Canadian GHG emissions had risen continuously and rapidly for two decades. Reductions in the long time frame were calculated as if Canada could start *now* and ignore all prior commitments.

The Harper government, as did those before it, continued to avoid any and all of the policies that might actually make a significant difference in outcomes: regulation of the GHGs from large emitters, carbon taxes, or a tough-minded national cap-and-trade system.

Canada Embarrasses Itself on the World Stage: Take Two

The Harper government soon realized that even if climate change denial and obfuscation were well received in Calgary and got mixed reviews in Ottawa, they did not play well on the international stage. The first Harper environment minister, Rona Ambrose, impressed almost no one, including the Canadian media, with her performance in Nairobi at the United Nations climate change summit in November 2006. Ambrose delivered a short speech that was widely seen as being totally inappropriate in an international setting.

Although she was less belligerent than Klein had been in Moscow, Ambrose again took Canadian domestic politics on a rough ride across the international stage. She acknowledged, as everyone knew, that Canada was far from meeting its Kyoto requirements, but mentioned no new initiatives aimed at achieving what the nation had agreed to or even at gaining ground within the Kyoto time frame. She did mention the possibility that mandatory limits might come into effect soon, but as it turned out no such limits were in place a year after her speech.

For the most part, though, Ambrose delivered a speech aimed at a Canadian domestic audience – speaking, of course, to an international conference uninterested in such matters. She placed all the blame for her country's ongoing failure on the previous government, asserting, "When Canada's new government assumed office this year, we found an unacceptable situation. We found that measures to address climate change by previous Canadian governments were insufficient and unaccountable."[11] She neglected to mention that her own party had denied the very existence of climate change and opposed all actions during the time that those measures were enacted. She also neglected to mention that her government had, upon assuming office, killed most of what had previously been enacted.

Steven Guilbault of Greenpeace, calling the speech "embarrassing," said: "It is obvious Minister Ambrose does not understand what it is to speak to the United Nations. She came here to wash her dirty laundry in front of the whole world." Instead of using an opportunity of a few minutes to assure the world that Canada would play a role in the international effort to fight climate change, Guilbault added, she "started pointing fingers."[12] The speech was widely panned, and the rising public visibility of climate change as an issue all but ensured that Ambrose's days as minister of the environment were numbered. She barely lasted into the new year.

As public concern regarding climate change deepened, the Harper government realized that denial was no longer a politically viable option – that climate change was, instead, an issue requiring finesse and at least the appearance of concern and action. Essentially, though, the Harper forces were still looking for ways to spin the issue into the oblivion they clearly preferred for it. Early in autumn 2006 Johanne Gélinas, the federal Commissioner of Environment and Sustainable Development, stated: "First and foremost, the government needs to clearly state how it intends to reconcile the need to reduce greenhouse gas emissions against expected growth in the oil and gas sector."[13] However, rather than even consider challenging the oil and gas sector on GHG emissions, the Harper government pushed for a new Clean Air Act focusing on smog. Many critics suspected that this effort was designed primarily to give the appearance of action on the environment and thereby placate the public's increased concerns.

Elizabeth May argued that while smog was a real problem, new legislation to combat the problem was unnecessary. As she put it:

> Canada's regulation of air quality lags behind other nations. Canada allows sulphur dioxide at concentrations of 115 parts per million, while the European Union allows 48 ppm and Australia permits 80. However, there is no need for new legislation to accomplish what an existing act could do very easily. The Canadian Environmental Protection Act (CEPA)

is the right tool. While a new government may enjoy the political and media boost of making new law, it is unnecessary, potentially costly, will impose years of delay and is likely to create a federal-provincial conflict for no benefit.

Regulating to reduce the precursors of smog, as well as to reduce serious neuro-toxic contamination with mercury, is possible within CEPA.[14]

In other words, adding new regulated substances to existing legislation or altering the level of control that already exists under that legislation would be faster, cheaper, and every bit as effective and decisive as passing a new Clean Air Act. The government's motives for introducing that act, one can only speculate, was to create the illusion that something was being done to "fix the problem with the air" that everyone was talking about. The hope was that some Canadians would pick up a vague sense that the government was acting on climate change.

On the contrary, though, as May pointed out, if the country did *not* act on climate change, problems with smog days were likely to get worse given the burning of more and more fossil fuels, even if, as a result of pollution-abatement technology, each litre burned emitted somewhat less of the contaminants that produce smog. Overall emissions of these substances might not be reduced much if at all, and "a failure to confront the climate crisis . . . will result in more extreme heat conditions. The more 30-degree days that Canadians experience, the more smog days will occur."[15] Heat, of course, is what converts smog precursors to smog.

A more charitable interpretation of the Conservative's proposal for a new Clean Air Act would be that it is not fair to criticize acting on smog as obfuscation. The government was free to *also* act on climate change, and that was a separate matter. Yes, the Harper government was seeking to get media attention and public credit for environmental action, but at least it was doing something on a front on which Canada had lagged other nations, including in some cases even the United States, for many years.[16]

This more charitable interpretation was undermined by the government itself only a few weeks later, when it announced further cutbacks to federal climate change programs. Some of the programs that had survived the April 2006 cuts were renewed for one year but were informed that they would not be renewed after that. These included programs based in Agriculture Canada: a model farms program to demonstrate how changed farming practices can remove atmospheric carbon, a program to involve farmers in GHG reduction, a shelterbelt that promoted tree-lined fields that would improve snow retention and cut erosion, and programs to manage manure and grow future fuel crops.

The opportunity to subsidize farmers was not being missed altogether, however, since the Harper government had announced an intent to encourage or require that 5 per cent of the fuel sold in Canada was to come from renewable sources, primarily ethanol, in much the same way as the United States was doing. The net climate change effect of this initiative, critics said, would be very small at best. Agriculture contributes about 10 per cent of all climate change effects, so programs to alter farm practices would seem to be in order – programs beyond the promotion of additional grain production, but several such initiatives were eliminated in April 2007.

Overall spending on the environment at Agriculture Canada was to drop from $331 million to $158.5 million by 2008–9. Natural Resources, which housed most climate change programs, was to be hit much harder. As a newspaper report on the cuts put it, "Overall, Natural Resources Canada estimates its total budget will drop from $1.47 billion this year to $1.04 billion two years from now and that the number of full-time employees in the department will drop from 4,456 to 4,154."[17] Cuts like this in the face of strong budget surpluses suggest that the motivation is not to "balance the budget."

Throughout 2006 the Harper government seemed to remain convinced that it did not need to do much to discourage energy consumption or to act comprehensively as a government with regard to climate change. The removal of Kyoto references from federal websites, acting on smog with great fanfare in lieu of acting on climate change, and systematic cuts to virtually all existing federal climate change initiatives – they all speak to a considerable indifference if not continuing hostility to the issue. The government seemed to be content to just push the GHG-reduction timeline into the future and to pretend that Kyoto had never happened.

Globe and Mail columnist Jeffrey Simpson, writing in October 2006, concluded that Harper was reading the party's polls of Canadian opinion when he selected this approach. Such polls probably suggested that concern regarding climate change was, as the pollsters say, thin – Canadians say that they care, but are really not intensely concerned and perhaps more than a little confused in many cases. As Simpson put it:

> Most citizens, [Harper] apparently believes, do not much understand the challenge of greenhouse-gas emissions and certainly won't change their lifestyles to do anything about them. . . . Global warming, he thinks, is no more or less important than other environmental concerns. In fact since carbon dioxide can't be seen the way pollutants that cause smog can be seen and felt, carbon dioxide emissions are actually less important, politically speaking.[18]

Thus Harper, given his political base in Alberta and in the oil industry, sought every way imaginable to finesse the issue, to avoid acting effectively by requiring strong actions on the part of the oil industry.

One of those attempts at finessing came straight from the Bush government's playbook. The government spoke of seeking reductions in greenhouse gas *intensity* – an artful twist that means fewer emissions per unit of product. In the case of the tar sands, this approach might mean somewhat lower GHG emissions per barrel produced, but a rapidly increasing number of barrels. Given that global warming is caused by the concentration of gases in the atmosphere, emissions of those gases must be reduced absolutely to lessen the problem. Emissions intensity counts for nothing; actual total emissions are all that matters to nature.

One other point about focusing policy and discourse on reducing the intensity of emissions needs to be made. If the federal government were to offer carbon emission credits for reducing GHG emissions intensity, tar sands operators could gain a significant income source by reducing emissions per barrel even as rising output radically increased their overall GHG emissions. At the same time Ontario Hydro, for example, might receive nothing for closing its coal-fired power plants because by some interpretations intensity of emissions are not reduced for those coal plants since they no longer exist.[19] That scenario would be perverse in policy terms, but not necessarily in political terms, especially when it comes to raising campaign funds and rewarding friends.

Intensity had previously been used by the Bush White House in its continuing attempts to confuse the public into thinking that something of consequence was being done to reduce GHG emissions. Indeed, the main climate change initiative of the Bush administration was an attempt to systematically control the public speech of scientists in the employ of the U.S. government.[20] Speaking of reducing emissions intensity has been part and parcel of this deliberate attempt at suppression and confusion. With the Harper government in power, the same expression found its way to Ottawa, though it did not catch on for long within everyday governmental discourse.[21]

The realization that denial was fading as an option fully jelled in December 2006, when Stéphane Dion somewhat unexpectedly assumed the Liberal leadership after running, against a formidable array of contenders, with a clear emphasis on climate change. With denial no longer plausible, the only hope was to turn to resisting Kyoto as the appropriate solution. Why? Because Kyoto mandates overall action on the part of Canada at a rate that is simply impossible without placing limits on tar sands expansion and providing tough actions on coal-fired power plants along with stronger versions of the sorts of spending initiatives that the Liberals, and to a lesser extent the Conservatives, have been prepared to support: providing incentives to Canadians to drive less, insulate their homes, change their appliances, and incentives to industries to gradually develop new technologies.

All federal governments thus far have been intent on keeping everything voluntary, for individual Canadians, for small businesses, for the auto industry, and above all for Alberta's oil patch. In the winter of 2006–7, however, this paramount goal was becoming an increasing challenge for the Harper government. With Dion in the

Liberal leadership chair, nature immediately seemed determined to show that being a climate change candidate was fortuitous.

The Winter That Wasn't

As if the ascendancy of Al Gore and Dion on the issue of climate change was not enough, the Harper government's position was further undermined by the 2006–7 winter-that-wasn't. As ski operators laid off employees and new winter coat sales plummeted, on January 4 Prime Minister Harper appointed John Baird, at the time head of the Treasury Board and purportedly a rising star, as the new minister of the environment. The party in power was making a seeming attempt to shift the rhetoric, to speak of fixing the problem – although still not within the Kyoto time frame.

Baird and Harper moved quickly in early 2007 to convey the new tone. On being appointed, Baird noted, "I grew up here in Ottawa, lived here my entire life, and I can't remember a winter where I didn't have to use my boots. I left my house without even a winter coat this morning. So that's obviously a huge concern."[22] From a socialist plot to a huge concern in a matter of months: the Conservatives were nothing if not adaptable. Dion was quick to acknowledge this shift and retorted regarding the Prime Minister's newfound concern: "It will not be difficult to do more than he did last year. Last year, he cut 90 per cent of the programs for environment and climate change."[23]

By late January 2007 everyone had noticed, and everyone was commenting. There really was no avoiding it even for those who had never heard of Gore (something that became harder to do as he won an Oscar and the Nobel Peace Prize in the same year). Every major newspaper featured commentaries on the unusual weather and linked it to climate change. *The Globe and Mail* featured front-page pictures of melting icecaps in Greenland, January golf, monster Australian wildfires, snow-free Swiss ski resorts, and trees in Stanley Park flattened from strong storms off the Pacific. Many people who were not at all active environmentalists were inviting their neighbours over to watch Gore's documentary.

Yet even in January 2007 Wente continued to argue that Gore was an irresponsible alarmist who "pushed the science way too far" – a point that she would presumably apply as well to other strong voices calling for action on the issue.[24] Wente was finally acknowledging that climate change just might be real and caused by humans, but continued to feel compelled to detail what science did not know and stress the many things where disagreement remained. Fair enough, but she was doing so primarily to identify those who push for action as "alarmists" or even as people who go "halfway around the world at someone else's expense" just "to yak" about global warming.[25]

Wente quoted Mark Jaccard, who has elsewhere done solid and interesting work on climate change policy tools, as saying: "Environmental activists are using climate change to wrap around their message about how they want humans to behave differently." What is missing from Wente's point and Jaccard's comments is an appreciation of just how hard it has been – and will continue to be – to get governments, any governments, and people – any but a very few people – to *actually* change anything from how it would otherwise have been (were there no such thing as climate change).

Yes, it is true that we do not know how fast change will come, or in any detail what forms that change will take through time, or, perhaps most important, how much we will need to reduce GHG emissions globally and how fast in order to avoid unacceptable harmful effects. Wente and others who are wary of action on climate change seem to have no sense of the level of inertia standing in the way of any reductions whatsoever, even in a generally well-intentioned and well-placed nation like Canada. With all of the alarmism and a scientific consensus, we have done nothing, *less than nothing*, in twenty years. It has been twenty years at least since the alarm was raised, and there has been *no* decrease in Canada's GHG emissions – only a continuing growth in output into the atmosphere.

The United States, the government of the most scientifically advanced nation on the planet and the largest GHG producer by far, is barely willing to even admit that there is a problem. China, the second-largest producer, whose GHG output is accelerating rapidly, for the most part does not consider climate change to be *its* problem. Most individuals do not imagine that they themselves could be expected to do much of anything, even if they think that *we* or the eternal *they* should do something about it. Wente expresses little concern about any of those issues; instead she worries that: "The global warming debate has become so polarized that it's impossible for even a reasonably well-informed person to figure out who or what to believe."[26]

Wente is not concerned that if that is so – or if people believe it when she says it is so – then even less action than has been taken in the past will be taken in the future. For me it is hard to even imagine how Canada could do much less in the future than it has done in the past. I'm not sure we could have increased GHG production more if we had actively tried to do so, short of torching the entire B.C. forest or building coal-fired power plants to replace Niagara Falls power. Overstating the concerns surrounding climate change could be problematic if those concerns turned out to be gross overestimates, but every indication (for example, the suddenly open Arctic waters of late 2007) is that the estimates regarding the possible rate of change have been too conservative, as scientists are typically and reasonably wont to be.

Nor does Wente appear to have a clue that slow GHG emission reductions can be accomplished relatively painlessly, with a minimum of economic disruption, but that rapid emergency reductions could be devastating. Why not err on the side of reducing energy demand sooner rather than later? Canada's Kyoto requirements could have been met had we tried, had we moved ahead with modest and effective programs from the mid-1990s or at least promptly in 1997 or 1998. Tar sands development might have been slowed for a few years and the emerging industry might have been ordered, following Kyoto, to sequester the emissions from all future plants beginning at some point in the then-near future.

Canadians were not alarmed by scientific overstatement or by scientific fact. Only the winter of 2006–7 got people's attention in any real way. If Canadians were paying attention in 2007–8 they will have noticed that people were collapsing from heat exhaustion in the Chicago marathon in October 2007 (not July). The problem is that this fact is alarming only if we appreciate that it will take many decades to alter global emissions and many more for that reduction to have much effect. To alter the human energy future in a way that only causes minor inconvenience and only hurts a few industries modestly will take a very long time, and all the "alarmists" are trying to do is to get us to *start* before that softer transformation becomes impossible, if it has not already.

We no longer have to argue about science, or study before and after pictures of global glaciers; all we need to do is spend more time out of doors and reflect a bit on how things were in winters gone by and what this might mean on a planetary scale over the medium to long term. Great certainty about exactly what will happen is not available, but does that mean that it makes sense to wait until palm trees are growing in Winnipeg and much of the planet is uninhabitable before we act – and just how, we might also ask, could we do so by that point?

Climate Change Meets the Holocaust

In May 2007 an odd incident in the continuing saga of climate change took place. Elizabeth May, by then the leader of the Green Party, quoted George Monbiot as saying that those who were not acting on climate change today were worse appeasers than Neville Chamberlain. Very quickly thereafter the leader of the Canadian Jewish Congress replied in a letter published in the *National Post*: "By comparing today's approach to the environment to pre-war approaches to the Nazis, Elizabeth May shows insensitivity to context and history." The exchange escalated rapidly in the hands of Harper, who said that May had "diminished the holocaust," though she had, of course, never mentioned the holocaust.[27]

Some newspaper columnists – Thomas Walkom and Rick Salutin, for example – defended May, noting that most of the politicians that commented and distanced themselves from May (or a leading member of their party) had also used a Neville Chamberlain analogy themselves, and so had Ariel Sharon in reference to George W. Bush, Ronald

Reagan, and Bill Clinton.[28] It appeared that neither the Canadian Jewish Congress nor Harper had been publicly offended by any of these previous references to Chamberlain.

Even more off the mark was *National Post* commentator Jonathan Kay in responding to Walkom. Kay wrote: "The reason people found May's comments so indefensible is that neither Stephen Harper – nor anyone who dissents from the Kyoto orthodoxy – seeks to kill a single soul."[29] Kay seemed to be confusing Chamberlain with Hitler. Chamberlain did not intend to kill anyone; he was seeking peace in his time, however foolishly. May asserted, perhaps inelegantly, that Prime Minister Harper (and President George W. Bush and Australian Prime Minister John Howard) were denying or foolishly ignoring a threat, a threat that carried very real dangers. She was making no claim that they were *intending* to kill anyone, or that they were Nazis: not even close.

All of this dissension – from some of the media assertions to the Canadian Jewish Congress to the responses of political figures – was a gift to Harper and to the cause of climate change action avoidance. As May apologized to those who were offended, the subject had shifted away from climate change and its dangers and from the ineffectiveness of the government's plans, such as they were, to deal with it. Classic political deflection prevailed as tender sensibilities were the order of the day (or week). Climate change itself was yet again all but lost in a fog of fretting.

In the conservative zones of Canada's news media, "Kyoto orthodoxy" became something to dread as much as the drowning of coastal cities or species loss or the loss of glacial runoff from the Himalayas or drought in Africa and Australia. Harper, it seemed, could be re-elected without providing a credible plan for climate change even if every other party had one, if only Canadians could be convinced that Kyoto was "political correctness" run wild. If only Canadians did not understand that Kyoto itself was a very modest compromise – a barely adequate beginning that Canada, a putative champion of internationalism, was rejecting outright every bit as thoroughly as Bush had – only just without coming right out and saying so.

The climate was clearly changing, and Canada continued to avoid any serious attempt at Kyoto compliance. Eventually the government would need to put the distractions aside and somehow face the issue, but how?

Harper Following the Winter That Wasn't

The Conservative Party program was, and is, to say the least, deeply flawed. At the same time, however, despite a career built on climate change denial, oil company sycophancy, and hard-core neo-conservatism – and a tenure as prime minister that began with the elimination of most of the modest climate change efforts that the Liberals had undertaken – Harper discovered that a melting planet was a hard sell, and as a result he shifted his rhetoric rather dramatically. By the end of the winter that never was, with Gore's Oscar and Nobel Peace Prize looming ever larger, Harper had packed away Kyoto as a socialist plot and took to admitting in public that the sea ice was disappearing from the Canadian Arctic.

Indeed, as time passed Harper realized that his problem with climate change was perhaps larger than he had anticipated even in January, when he first sought to get his party's rhetoric on the issue under control. His approach, by the spring and summer of 2007, was to communicate that Kyoto, the treaty he had badmouthed for a decade, was now hopelessly out of reach. Taking this tack, he could do little that was substantive about it and still get re-elected. Whatever else one might say about Harper, he could not be accused of an unwillingness to adapt, at least in terms of rhetoric: in 2007 he was verbally a veritable climate change chameleon, especially on the global stage.

In late April Baird issued an announcement regarding actual government climate change policy. Baird spoke of reduction targets of 60 to 70 per cent by 2050 and 20 per cent by 2020 – though he did not make many comparisons to 1990 as a base year – and it is the base year on which the Kyoto agreement rests. In that light the 2020 target is 2 per cent *above* 1990 levels. That is, Canada would be 8 per cent *above* Kyoto requirements eight years *after* the final deadline for meeting those requirements; and that would only be the case if Canada actually met those targets.

One had to read the fine print, however, and the Pembina Institute did just that. The Harper team asserted that the growth in Canada's GHG emissions would peak somewhere between 2010 and 2012, but, according to Matthew Bramley of the Pembina Institute, the government "provided *no explanation* as to how it expects to meet its target for national emissions to peak during 2010–2012. Without measures additional to those the government has announced to date, the short term target can only be met if there is an unexpected and dramatic slowing of the business-as-usual increase in emissions."[30] If the past twenty years is any guide, saying that it would be nice if this were the outcome is not likely to produce that outcome.

The government did mention the possibility of a "regulatory framework" but avoided setting firm regulations. Moreover, even the *asserted* goal for 2020 for heavy industry is only 18 per cent below 2006 levels (12 per cent above 1990). The Pembina review points out that the government's goals are unlikely to be met for a long list of reasons, including: the targets are expressed in terms of intensity, not emissions; and industry targets are heavily backloaded towards the period just before 2020 and are thereby unlikely to be met given that progress will be reviewed in 2012 and little will be done by that point. Moreover, in the government's plan the energy sector is not required to put in enough money to pay for carbon capture and storage, and new oil sands facilities are treated more leniently than are established facilities, when the opposite makes far more sense technically and economically.

More than that, the industrial sector is responsible for more than 45 per cent of all emissions, and even if it meets its implausible targets – implausible given the lack of a firm legal requirement to do so – all other sectors must do better than industrial emitters, and few programs are in place to suggest that this will happen. Again, that kind of progress has certainly not happened over the preceding twenty years. Finally, if they get credits for "early actions" already instituted or if they pay into a technology fund that will be used to create reductions, industries can meet targets without actually reducing even in terms of "intensity." The technology fund option means that if these technologies are adopted, as Pembina points out, any reductions achieved will be counted twice.

Harper faced further pressure to adapt when, in June, the three opposition parties in the Parliament united (a singular event indeed) to pass the Kyoto Protocol Implementation Act over the objections of the minority government. The act forced the Harper government to offer new climate change policies by August, while also requiring that those policies be assessed for effectiveness by the National Round Table on the Environment and Economy within thirty days. In August, as the deadline approached, the Harper government essentially recycled the policies that Baird had announced in April.

The NRTEE assessment, duly released in a report in late September, was highly critical of Harper's climate plans. It saw the government's nine programs as inadequate to the task of complying with Kyoto. The report, according to environment reporter Peter Gorrie, "accused the Conservative government of using 'systematic' exaggeration and 'double accounting'; of 'not accurately reflecting' emissions reductions, and of using 'important inconsistencies' and of having 'overestimated' reductions to produce false conclusions about the effectiveness of its plan."[31]

The NRTEE concluded that "of the nine federal climate change programs it studied, the government had exaggerated the benefits of three and failed to produce sufficient information to support the other six."[32] Environmental groups, including Friends of the Earth, responded by taking the government to court, asserting that it had not produced a plan to even attempt compliance with the Kyoto treaty as required by the Kyoto Protocol Implementation Act.

The NRTEE's report was, indeed, very similar to that of the Pembina Institute regarding the earlier (and almost identical) version of Harper government policy, but the government continued on as if no one had noticed. Indeed, it would make the same claims about 2020 and 2050 in the speech from the throne a few weeks later.

In that October 2007 throne speech the Harper government even acknowledged: "Threats to our environment are a clear and present danger that now confronts governments around the world. This is nowhere more evident than in the growing challenge of climate change."[33] Yet in the same speech Harper also made it clear that Canada's Kyoto commitment was dead, and he cast all the blame for that sorry result on the Liberal governments whose every action he and his party had resisted tooth and nail.

Regarding Kyoto, the throne speech said: "At the end of 2005, Canada's greenhouse gas emissions were 33 percent above the Kyoto commitment. It is now widely understood that, because of inaction on greenhouse gases over the last decade, Canada's emissions cannot be brought to the level under the Kyoto Protocol within the compliance period, which begins on January 1, 2008, just 77 days from now."[34] The speech did not mention that the actual compliance deadline was more than four years away, nor did it mention any new initiatives to make at least a significant effort at compliance within that time frame. The government's targets were all safely beyond its term of office, in 2020 and 2050.

The government assertions remained vague and its policies unchanged despite the negative assessments of the Pembina Institute, the opposition parties, and all major environmental organizations. All of this, too, came despite the requirement passed by Parliament in June to issue policies to comply with Kyoto. The October throne speech was in effect a case of standing pat for the second time as the government's climate change rhetoric on the global stage continued to build.

In September Harper had spoken at the Summit meeting of the Asia-Pacific Economic Cooperation (APEC) organization in Sydney, Australia, where he stunningly asserted: "Canada wants to be a world leader in the fight against climate change and in the development of clean energy."[35] Press coverage of that speech noted: "The Prime Minister, who five years ago dismissed the science that linked emissions to climate change as tentative and contradictory, said 'the weight of scientific evidence holds that our atmosphere is getting hotter and that human activity is a significant contributor.'"[36] His rhetoric had reversed so thoroughly that one could almost hear the echo of an invisible drill sergeant shouting, "About face!"

The Conservative policies had apparently not heard the drill sergeant, nor did, or would, Harper deliver a similar speech in visits to a place like Calgary. Indeed, Harper still held that the Kyoto targets were impossible and that Canada was looking for GHG strategies that were "comprehensive, practical and realistic." Practical and realistic to a neo-conservative mean placing no undue pressures on big industries, and "undue pressures" refers, it seems, to anything that those industries believe they should not have to face. Even for a skilled political leader like Harper, however, it is difficult to do a full reversal without actually changing direction.

From April to October little had changed, and neither policy implementation details nor any response to criticisms ensued. The government's overarching approach to the issue seemed apparent: say the right things about the problem and set reasonable long-term goals, but do as little as possible, as slowly as possible, without setting any firm rules regarding emissions.

Fortuitously for the Harper government, though, political events and circumstances late in 2007 conspired to prevent the opposition from again acting in a united fashion. In particular, three by-elections in Quebec gave the Conservatives an opportunity to put a stake through the heart of Kyoto. The Liberals lost ground in Quebec, with one of three by-election seats going to each of the NDP, the Conservatives, and the Bloc. Shortly afterwards, internal Liberal criticisms of Dion as party leader found their way into the press. The once famously united "natural governing party" was again unable to keep its petty spats out of the public eye.

With the Liberal Party both divided internally and noticeably weak in its fundraising capacities, Harper decided to play hardball. He made the contents of the throne speech a matter of confidence, daring the Liberals to vote against him and force an election. The Liberals bristled at several aspects of the speech, including the assertions regarding Kyoto, but opted to abstain on votes that would result in an election if the Conservatives were to lose. With that the new government policy was to abandon Kyoto, toss off policies that did not meet the legal requirements of the Kyoto Protocol Implementation Act, and instead assert targets for 2020 and 2050 without implementing any policies that would convincingly advance Canada to those targets (targets that were short of Kyoto compliance *even in 2020*).

Thus, as the 2012 Canadian GHG targets joined previous targets and deadlines in the dustbin of history, 2020 and 2050 became Canada's new target dates for *really doing something*. In the meantime tar sands expansion plans continue to unfold, with no delay of additional new plants until they can meet emissions requirements tough enough that the efforts of municipalities, other firms, and Canadians would not be overwhelmed by the scale of the emissions from Northern Alberta alone. Also left off the agenda was any inquiry into how much oil Canada should export, on what terms, and under what conditions (with regard to water quality, for example).

Somehow we have arrived at a world in which concerns about the physical systems and ecosystems that sustain life on earth are not "practical and realistic." Realism above all acknowledges and accommodates political and economic power, even if those with power do not consider the death of coral reefs and the loss of drinking water from glacial runoff for millions of humans a high priority. Perpetually acceding to the preferences of the rich and powerful creates a world in which everyone waits for everyone else to take responsibility for their actions.

Again, as *The Globe and Mail* account of Harper's speech noted, "Critics of harsh measures to fight climate change have pointed out that Canada produces only 2 per cent of global greenhouse gases and no real reductions will occur until the large producers like China adopt strict environmental policies."[37] Climate change naysayers and entrenched interests in other nations have, of course, made the same "what we do is not important" argument about emissions originating in their countries. More important, China's per capita emissions are only a small fraction of those of Canadians or Americans, or even Europeans. The China excuse is the very one offered by the United States (which cannot dismiss its emissions as only a small percentage of the global total) regarding why it has not signed Kyoto. Again, everyone who wants not to do anything points to someone else.

Practicality and realism, of all things, all but dictate that the same looking to everyone else to go first can now occur within Canada. Individuals may do the painless things – put in efficient light bulbs and read the EnerGuide labels on the new appliance with some care – but it is a real moral challenge to do the tough things when prime minister after prime minister is at least implicitly clear that large industries will not be pushed hard to do what is tough for them. For some individuals the tough things include paying $8,000 extra for a new hybrid vehicle or for replacing all the windows in a home, and for others they include not taking a winter holiday in the Caribbean or Florida even if it's affordable.

Especially hard to change, however, are the habits of a lifetime. Giving up driving several hours each way each weekend to a cottage that you remember from your childhood is very hard to do, and few would voluntarily do it for any global cause, however noble or urgent. Nor is it easy to give up on long-standing dreams to travel to distant corners of the world after you retire (or before you settle down to earn a living and raise a family). It is also very difficult to imagine not seeking the comforts of air conditioning or other seemingly small luxuries, especially, perhaps, if you are old or ill. Hard choices may be unavoidable as part of a comprehensive global solution, though perhaps a little imagination will reduce the number of such choices that must be made.

Some changed habits are, of course, good for us in the long run (walking more, or bicycling, for example), and some things that seem to cost more in the short term will result in savings in the longer term. Nonetheless, people will only change their behaviours if the need is obvious, and if the burdens of change are fairly shared, and if the cause doesn't seem all but hopeless – for it is all too easy for humans to lapse into a state of hopelessness and all too easy to accept immediate pleasure and convenience over long-term gain.

Canadians will make changes, maybe even significant ones, if they see others making them all over the world as well as in their own neighbourhoods and social circles. That is how recycling took hold a few decades ago. Canadians might even rearrange their vacation days and take longer weekends at the cottage every other week or once a month. But letting the oil industry rapidly increase its GHG emissions to accelerate energy exports and allowing coal-fired power plants to stand pat send exactly the wrong signal to anyone taking time to consider what they might do for the cause.

We have to hope that the deeper sacrifices are not necessary when it comes to avoiding the serious effects of climate change. No political leaders and few people would imagine that they will be, but if Canada's energy exports grow at the rate they are projected to grow and GHG emissions grow anything like proportionately, such sacrifices by Canadians would become both more environmentally necessary and less possible politically. Then even the easy steps would raise resentment and be dismissed as hopeless idealism in the face of global forces over which we have no control.

In dealing with climate change the ability of a government like the one led by Harper is inherently limited because neo-conservatives instinctively do not trust government and doubt the possibility or desirability of collective action. This is only in part because they are wary of taking actions that might undermine their political appeal to business interests, interests that provide a significant proportion of their campaign funds. It is also because they genuinely believe that the market can more efficiently allocate resources to get things done. That belief is, unfortunately, only a partial truth.

Conservatives and many others just do not see that competition does not efficiently allocate goods and services when the corporate interests involved are large. Economic interests will fight to avoid increasing costs or reducing consumption of the products that they produce. They will act very effectively through the media within the political arena to block government from acting or the consuming public from reducing consumption. They will advertise to create demand. They will lobby to avoid any and all regulations. They will work with other corporations to block actions that would require, for example, the production of more fuel-efficient vehicles (or subsidies to transit use). They will work hard to create and maintain dominance in the marketplace through *political* means.

Neo-conservatives believe that what they are doing is protecting the "free" marketplace from government interference, but corporations are working to use government to organize the marketplace in their favour. Corporations sometimes think outside the time frame of today's market, or the next quarter's results (or perhaps in terms of the life of a new factory), but only rarely. Most corporate executives do not have the time or the inclination to think very hard about issues beyond their balance sheets, or very often regarding the long term.

Climate change is thus an issue largely beyond the ken of most conservatives and the inclinations of many (though decidedly not all) corporate leaders. As such it is just the sort of issue that requires strong government action, especially in relation to the energy sector. Government must alter the equation beyond a competition between doing what's right for shareholders in the short term and doing what's right for the planet in the long term (and perhaps for public relations). In that contest the balance sheet will almost always win.

In this regard a government can tip the scale in at least four ways: regulations that limit actual emissions rather than putting mild and flexible limits on intensity, taxation of GHG emissions, incentives to key industries to reduce emissions, or establishment of a cap-and-trade system – the creation of a market in GHG reductions.

The Harper government offered incentives to industry, but they are far from adequate in and of themselves to bring about significant change. It eliminated some incentives to households. It claimed that it was regulating industry, but the process it introduced was fraught with loopholes. The Harper government would not touch carbon taxes – leaving only the possibility of cap-and-trade, which is a system that could help to lower the cost of emission reductions by industry, but also one that needs to be carefully designed – something that should have been done decades ago. The Harper government's plans on this front are hardly sure to be effective, if they even ever happen.

The Harper government's policy regarding emissions trading has been to "explore opportunities" to link into a North American GHG emissions trading system. Such a system unfortunately does not yet exist, nor is there such a system in place in either the United States or

Mexico separately. There is an established system in Europe, but the Canadian government said nothing about the possibility of joining in an effort in this regard where one actually exists.

No such system is anything like imminent in the United States. There is no assurance that it will ever be. More than that, to be effective such systems need to be established in such a way that total emissions are continuously reduced, that outcomes are monitored, and that Canadian corporations are not simply able to avoid action by spending a small amount of money ineffectively in low-cost jurisdictions. Whether the Harper government's claims regarding emissions trading will ever amount to anything is utterly unknown, but talking about the possibility does convey the appearance of motion, and that would appear to be the Conservative goal.

Thus the Harper government's climate change plan has not been a plan in the usual sense of the word. The plan is to introduce long-term targets, but not to worry a great deal about how those targets might be achieved. Part two of the plan is to talk a lot about climate change and Canada's intentions on the global stage, and on that stage to say all of the right things. There is no part three.

Finally here is the question of legality. Canada signed the Kyoto Protocol, and treaties are law. Inaction on the matter by a series of governments is every bit as illegal as the Liberal's sponsorship scandals were, and in the end more important to Canada's reputation and the well-being of the nation and the world.

For twenty years all manner of policy processes have been undertaken in Canada, and in twenty years nothing of consequence has been done. Canada's total greenhouse gas emissions have never been reduced in any year – not once thus far. Nor has any government ever actually required anything of any corporation regarding emissions. Some corporations have made efforts, but many have not bothered to do much of anything. The very largest sources in Alberta's oil patch have grown like Topsy.

This outcome can be explained at another level – one that goes beyond the day-to-day foibles and troubles of the succession of Canadian political leaders that have dealt with the problem. Canada has the capacity to be a world leader on climate change action. Canada has long-term fossil resources, but it also has bountiful non-fossil energy possibilities and significant room for energy efficiency gains. Canada has vast wealth, technological capability, and energy production experience. Moreover, the export of even a fraction of our energy reserves and capabilities generates sufficient money to finance a domestic energy use and alternative energy revolution. What we might learn in such a transformation (regarding carbon sequestration or biofuel production, for example) could also be widely exported.

What has prevented effective action thus far are shortcomings in our political system, rather than material or economic limitations, or the absence of a desire on the part of Canadians to do the right thing – to live up to our global responsibilities as a prosperous and privileged nation. Indeed, most Canadians want very much to play a positive role on the global stage in this matter and in all others. But at least three shortcomings notably act to prevent us from playing that role with regard to global warming.

First, in Canada the greater share of constitutional authority regarding natural resources rests with the provinces. A very large proportion of our fossil energy reserves, and the seeming capacity to develop them at a rate that precludes meeting our international climate change obligations, rests with one province in particular. Or at least the leaders of that province assume that that is the case – and no one has yet disobliged them of this belief.

Second, the Canadian political party system is highly fragmented. Majority federal governments are increasingly hard to come by. There are now five viable federal parties, four of which favour strong, or at least greater, climate change action. The four parties together command a very large majority of the popular vote, but given that they are polarized on the question of Quebec, that they disagree about many other matters, and that there is no tradition of coalition government in Canada, they cannot easily form a unified government – one that could stand up to provincial claims of authority on matters related to energy and climate change policy.

Third, Canada as a nation is deeply integrated within the North

American economy, and many Canadians believe that it has limited power in that setting. Historically Canada has indeed had little influence in consequential matters of economic significance. Our exports to the United States are of greater consequence to us than U.S. exports of comparable size to Canada are to the United States. This is a function of the relative size of our economies, but it could be argued that present circumstances are significantly different. Our exports are increasingly dominated by energy, and politically secure supplies of energy are increasingly a product unlike any other.

As a result, then, Canada continues to face great challenges in mounting effective national climate change action.

Six

Frozen Governance in a Melting World

ALTHOUGH THE CHRÉTIEN Liberals eventually summoned the courage to sign and ratify Kyoto, no Canadian government since has acted effectively on that commitment. Canada hosts global conferences and even Stephen Harper, as prime minister, learned how to say the right things in international settings, at least in public. But effective action continues to just slip away year after year and decade after decade.

In the period following the 1988 Toronto conference and running through the signing of the Kyoto Protocol, Canada performed less well than did the United States. Between 1990 and 2000 the United States' emissions increased by 14 per cent and Canada's emissions increased by 20 per cent.[1] With limited action the norm before and after ratification, Canadian governments have in effect turned their backs on one of the most important treaties that the country has ever signed. In twenty years Canada has not mustered the political will to issue *any* climate change regulations other than the announced ban in April 2007 on the sale of incandescent bulbs – to take place by 2012, about when the industry might have stopped making them in any case (given that major retailers like Home Depot are phasing out sales). Nor has any government initiated a carbon tax or established a working cap-and-trade system.

An exclusive dependence on discussions, government spending, and voluntary actions should be identified for what it is: evidence of policy failure. Canada's failure to act effectively with regard to climate change is symptomatic of a fractured and flawed democracy. As a nation Canada functions very well in many ways, but it is highly prone to the pull of regional power and to weak and ineffective national governments.

Canadians as a whole are well intentioned in both international affairs and environmental policy. These attributes are almost as much a part of being Canadian as is our health-care system or hockey. Endless polls distinguish us from Americans in our approach to international relations, and Canadians on the whole are nearly as environmentally oriented as Europeans.[2] But public attitudes do not necessarily lead to appropriate action.

Political history, national linkages, and institutions count for a great deal in political and policy outcomes – much more, perhaps, than do public attitudes at any particular time. In the case of Canada and climate change a number of interplaying factors have pushed our nation in several directions at once. A couple in particular have made it difficult for even well-meaning governments to act effectively. The first is a party system and a parliament caught somewhere between a two-party system and a multiparty, European-style governance. The second is that we are a semi-fractured nation, and as a result the Government of Canada is prone to domination by the governments of the larger, and especially the more fractious, provinces. Quebec is Quebec, perpetually ambivalent with regard to being a part of Canada, and certain Western provinces, especially Alberta, are constantly on the watch for incursions on or limits to their newfound power and wealth.

Yet another factor might even be the most important: Canada's habitual role within an integrated North American economy. Since Confederation Canadian governments have continued to think like junior partners – first of Great Britain and then of the United States. Obviously we remain in that position in terms of our population size and military capacity, but I would argue that on the matter of climate change we have good reason, at this particular moment in history, to see our way past that traditional role. The world has changed, and perhaps Canada as a whole has not fully appreciated how much it has changed. Canada has much more leverage on the issue of climate change than it might at first glance seem, if only we are prepared to act.

A Distinctive Party System

Most Canadians do not appreciate how unusual our system of political parties is. Canada has evolved into a five-party system with frequent minority governments, even though we have an electoral system (single-member constituencies, with first-past-the-post tallies) designed to force majority government. Party systems almost never evolve in this way without an alteration of their electoral systems. Third parties in the United States, for example, are routinely crushed, hardly cause a ripple, rarely last more than an election or two.

Great Britain has a similar electoral arrangement, and it has had a two-and-a-half party system most of the time without a strong tendency to multiple parties. Most multiparty systems are linked with some variation on proportional representation – a system wherein parties can establish themselves in the legislature if they get even a limited vote, thinly distributed (which is much easier to do than getting a plurality in one or more ridings). In 2007 Ontario voters rejected the possibility of moving to a partial system of proportional representation for a limited number of seats.

Going from two or three to five viable parties without a change in electoral systems is exceptional, but it is also not without special problems. When it comes to important issues Canadian election results can seriously distort the will of the electoral majority – which is the case, as we've seen, on the question of climate change, given that four of Canada's five political parties, representing 60 to 65 per cent of the electorate, and holding among them a majority of seats, want stronger government action on that front.

The Liberals signed Kyoto. The man chosen as leader of the Liberal Party in late 2006 won the job against other seemingly stronger contenders in part because he spoke forcefully and clearly on this issue. The leader of the New Democratic Party has been an environmental activist for decades. Climate change was his signature issue when he served in municipal government and as a leader of the Federation of Canadian Municipalities. He retrofitted his own older Toronto home at great expense to a very high environmental standard. He pioneered a program that would encourage the renovation of a high proportion of Canadian homes. In many ways he and his party out-green the Greens, or at least attempt to do so. Despite this the Greens burst out of nowhere in recent years, sometimes polling above 10 per cent in popularity and showing consistent, strong growth in support in large part because of their emphasis on climate change.

That is a strong set of credentials on climate change for these three different Canadian political parties. Yet although few Canadians outside of Quebec appreciate it, the Bloc Québécois also takes a strong stance on climate change. The issue features prominently on the party's website – indeed, more prominently than does Quebec independence or any other issue. The Bloc does regionalize the issue, emphasizing Quebec's lower per capita emissions and singling out Alberta as a significant source of the problem that Canada faces as a whole. But it also advocates a variety of strong policies to deal with GHG emissions.

These tendencies all point to a structural problem in Canada's electoral system. An overwhelming majority of Canadians want strong action on climate change. Recent elections have produced a strong majority vote for parties that want to see action, yet those parties have been unable to form a government. One reason for this outcome is that Canada has a limited history of and experience with coalition governments. Another is that the Green Party vote has thus far been spread too thinly to win even one seat under the rules of the electoral game. If the rules changed to a variation on proportional representation, voting patterns might also change and the Greens could elect at least several members. Given the 2007 decisive defeat in Ontario of the referendum on Mixed Member Proportional voting, however, such a change – or even the opportunity to vote on such a change – is not likely to come soon at the federal level.

In any case the various parties, and most dramatically the Liberals and the Bloc, have a deep history of distrust, even fear and loathing. It is difficult to imagine any circumstances in which the Liberals and Bloc Québécois could co-operate, let alone form a government. Nor would a Liberal-NDP governing coalition be easily achieved, though in this case there is at least a history of negotiated co-operation of some durability springing from the days of David Lewis and Pierre Trudeau in the 1970s, when among other things the NDP managed to have Judge Thomas Berger named to head up the inquiry on the most contentious issue of the day, the Mackenzie Valley Pipeline.

Given that today's five parties are becoming ever more well established, the parties of the centre-left should be more open to the idea of establishing coalition governments. At the same time, Canadians should be exploring changes in the electoral system to help to ensure that dispersed electoral support has a better chance of influencing electoral outcomes, ideally in all regions. Given the existing pattern of party voting preferences, Canada should be able to elect Liberals, NDPers, and Greens even in Alberta, and more Conservatives in the Maritimes and Quebec. Generally the number of seats should represent prevailing electoral preferences.

Other reform possibilities include mixed electoral systems, which allow both for some members representing constituencies and others to be elected on a province-wide basis. Also possible, well short of proportional representation and its complications, are preferred-rank ballots that allow voters to select their first, second, and third choices for candidates and have their succeeding choices counted if their preferred party was the least successful (until any one candidate in the riding received 50 per cent).

Under our electoral arrangements now, the best outcome in terms of effective action on climate change would be a three-party coalition government, with tacit support from the Bloc. The Greens and NDP might be able to push the Liberals to silence some of the dissent within their party and take a stronger stance. For this to happen we would need a four-party majority, something that would most likely require the election of more NDP representatives and several Greens as well – not an impossible outcome, but one that would not be easy without the simultaneous loss of some Liberal seats. The dilemma on the climate change issue is typical of the problems that the country faces with its electoral politics generally.

A Provincial Nation

The second challenge facing Canada in getting effective action on climate change is, to put it bluntly, the excessive power of provincial governments. Canada exists within a wholly integrated North American economy dominated by the United States. Global economic integration has resulted in a world in which the rules of doing business must be set globally so that Canadians, and everyone else, are not forced to accept diminishing environmental protection and social benefits in order to protect employment opportunities. Many of today's pressing problems – from terrorism and the risk of disease pandemics to climate change – can only be solved on a global basis.

Yet in Canada provincial governments have the power in many cases to block or limit effective national action on the international stage. On the issue of climate change, every Canadian national government must perpetually look over its shoulder even when only one province – even one province whose government is dominated by one industry – stands firm against effective action. This is the case even if most Canadians, and indeed most people in the world, not only want but also require collective action for the well-being of future generations the world over.

Canada's provincial governments cling to their prerogatives as if nothing has changed in a century or more. Some provinces still hold attitudes associated with "Western alienation" many decades after the grievances in question were at all meaningful. Given the changes that have taken place – global economic integration and the rise in power in the Canadian West, especially Alberta – the Canadian Constitution places the capacity of our nation to play an effective role in the world regarding climate change in the hands of one or a small number of provincial governments.

This is not to say that provincial governments should not play a role in international affairs. The question, however, is this: Under what circumstances should the federal government have the capacity to override provincial power in the national interest? The framers of the Constitution clearly understood the dangers inherent in excessive provincial power. The Constitution grants the federal government the power to act in the interest of "peace, order, and good government."

Good government in Canada in a contemporary context would see our national government act with other nations to prevent the melting of glaciers that millions of people around the world depend on for fresh water. It would see Ottawa take action with other nations on the death of the coral reefs that provide a high proportion of global fish stocks, on the drowning of Bangladesh, Southern Florida, and New Orleans, and on the spreading of deserts and political instability through much of Africa. The interests of peace and order would also imply the same action.

Arguably, order within Canada itself depends on effective Canadian and global action on climate change. Should profits in Alberta's oil patch and their influence on Alberta governments control all aspects of resource development in that province – even if the result is the destruction of the polar bear habitat and permafrost in Canada's Arctic? Even if it means that the money spent by well-meaning Canadians everywhere to upgrade their appliances and buy more fuel-efficient vehicles will be overwhelmed by one export-oriented industry operating in one Canadian province? Even if it means that Canada is no longer trusted and respected in the wider world for having broken its word on doing its part – let alone taking a decisive lead – on an obviously serious global problem? What *are* the limits of provincial power?

Ironically, the Canadian Constitution, in direct response to the U.S. Civil War, was designed to give the national government general precedence over our provincial governments (while the U.S. Constitution was written to avoid undue concentrations of power anywhere; it thereby favoured decentralization). Through time the United States, as its national economy grew more integrated and its interaction with the rest of the world increased, centralized a larger share of governmental authority in the national government. Canada, in contrast, allowed the federal government to become so relatively weak that seemingly nothing of consequence can be done without a federal-provincial conference. After all, in 2002, after five years of consultation, it was Alberta's Klein who accused the federal government of acting on climate change "without consultation."[3] But, in fact, the federal government had not acted decisively to reduce GHG emissions in Alberta or anywhere else. Rather, it had merely ratified, after endless years of delay, an agreement with other nations that it *would* act. The federal government had yet to take a decisive step with regard to climate change and never did act decisively even after signing and ratifying the agreement that threw Klein into regular public tantrums.

Continentalism as a Climate Change Lever

The third item on our list cuts both ways. Canada's deep integration within a continental economy dominated by the United States often makes daringly autonomous action deeply challenging. But at the same time it means that if Canada were indeed able to summon the political wherewithal to act decisively on climate change, the effect of those actions would probably be greatly magnified.

Given our energy resources, and the desire among ordinary Americans for their nation to play a different role in the world after eight years of the Bush government, Canada could conceivably help to tip the balance on climate change within the United States. As they say: pie in the sky, but still worth a try.[4] Given that a concerted campaign on acid precipitation in the 1980s turned the United States around on that issue, what might a well-timed effort on climate change do? Especially when most of the rest of the world and many, many Americans are pushing in the same direction?

Within the United States, the view of the oil companies, the auto industry, and the Bush administration has already lost ground. Gore has done a great deal to move the issue onto the U.S. political agenda, but he is far from alone in pushing things forward. The mayors of most large- and medium-sized cities are committed to taking action.[5] Arnold Schwartzenegger, as the Republican governor of California, has taken significant recent initiatives – even visiting Ontario in 2007 to advance the issue.[6] Indeed, one California initiative could ultimately limit the sales of oil from the tar sands in the state on the grounds that from well to wheels it is a far more carbon-intensive fuel.[7] The governments of most Northeastern states have also taken strong action.[8] Ed Rendell, the governor of Pennsylvania, has co-operated with the Apollo Alliance, a labour-environmentalist coalition looking to create a wide array of construction and industrial jobs in the resolution of climate change concerns.[9] Indeed, it can be argued that acting on climate change could reduce America's massive trade deficit and create domestic employment opportunities.

The United States is thus far from being a monolith of climate change opposition and denial. By 2008 President Bush and the Republican Party were as out of step with American opinion on this issue as they were on any issue save perhaps the occupation of Iraq. Opinion studies indicate that in general the U.S. governing elites (especially Republican Senators and staffers) dramatically lag behind public opinion on this issue.[10]

This context creates a chance for Canada to tip the balance of U.S. opinion and help to provoke a shift in that country on climate change. To do so, however, Canada, must act decisively, demonstrably, and forcibly. Such opportunities do not appear frequently, and this may be a rare chance akin to the one that occurred twenty years ago when Canada was the tail that wagged the North American dog on the leading environmental issue of that day: acid rain.

Canada applied diplomatic pressure regarding U.S. acid emissions from coal-fired power plants in Illinois, Indiana, and Ohio – emissions that found their way north and east and fell on Canadian lakes and rivers from Ontario to Nova Scotia. Canadian urgings complemented internal pressures from environmental organizations and from the states of Vermont, New Hampshire, New York, and Maine, where similar effects to those felt in Canada were being imposed by emissions from the same Midwestern U.S. sources.

In the case of acid rain Canada mustered the gumption to go beyond the usual diplomatic niceties. Canadian environmental organizations opened offices in Washington, D.C., and they lobbied U.S. legislators directly. They worked at getting media coverage, worked co-operatively with U.S. environmental organizations, and learned their way around the Byzantine inner workings of power in Washington. The campaign was one of the most challenging and successful efforts in the history of the Canadian environmental movement. Another key to success in the case of acid rain was that Canada had already undertaken solid initiatives of its own before pushing the United States to come along.

On climate change the undertaking is perhaps more challenging, but in this case Canada has a weapon at its disposal – secure energy sales – that it did not have regarding acid rain. The first step is simply to demonstrate that a rich North American nation can deal effectively with GHG emissions. It would require a significant turnaround, but Canada could impose costs on energy extraction in Canada, costs that the industry will fight against, but could easily bear. Canada could also show that with incentives North Americans will buy more fuel-efficient vehicles and appliances, that they will take up incentives on energy-efficient home renovations, and that people in urban areas will get out of their cars and use public transit.

Make no mistake, though, the decisions that must be made are not easy decisions. The path to be taken – on the issue of the tar sands, for instance – is probably more a political than a technical challenge. Canada needs political leaders who can think long-term, at least on this one issue. Policy is ordinarily made to maximize economic growth in the short to medium term. Policy is made to satisfy powerful interests and powerful neighbouring nations. Slowing or delaying tar sands expansion for even a few years would be a very strong measure for any national government. There would be money to be made that, in the minds of those who would potentially make it, would be "left on the table." For them, the prospect of making even more money five or ten years down the road, when energy prices will almost certainly be higher, would not make it easier. Nonetheless, Canada should have a national debate – that is, all Canadians should have a voice on these issues.

Effective action on climate change motivates a wide array of opposing forces – a much wider array, with much deeper pockets, than was the case with acid rain. Unlimited political funds will flow to the party that will look away from unrestrained expansion. A constitutional struggle and very serious economic threats from the Americans will greet any Canadian leaders who seek to cap the rapid and unlimited expansion of oil output. Yet at the same time Canada's capacity to supply a share of U.S. energy needs from a politically secure setting is a lever that is potentially of considerable consequence in the right political setting and in the right hands. Canada *could* discreetly, and out of public view, apply that lever to induce the United States to sign on to a reasonable international accord in the next round of climate change negotiations. Doing so would, of course, be unlike anything Canada has ever previously attempted. The audacity involved would be highly un-Canadian.

But it really does come down to this: Is Canada an independent nation with a different approach to the world than the United States, or is it not? Is Canada not a little bit different at least? Can Canada figure out how to effectively place our weight on the side of those Americans who already understand that salvaging their country's reputation requires a dramatic shift in policy? Could Canada make it easier for *those Americans* to prevail in the coming years, beginning with the issue of climate change and energy transition?

The United States' energy problems run much deeper than ours. The country's total energy need is ten times ours, and its conventional oil and gas reserves are in steep decline. It has oil shale, but no certainty that the resource can be extracted economically, and there is not sufficient water in the dry West where it is located to extract a great deal of it in what is likely to be a water-intensive process. Given the U.S. demand for oil, the Alaskan reserves are of minimal consequence. The proportion of oil imported into the United States has been rising for decades, despite glorious campaigns for energy independence dating back to presidents Carter and Nixon.

Energy independence in the United States is not an option at anything near to the country's present levels of energy use or with its present array of supply sources – but Canada can by no means just sit back and be smug about that reality. Disentangling our two economies might be technically possible, but it could not be done quickly and, moreover, the political fallout from trying to achieve greater autarky (economic autonomy) would be spectacular. Even after the adjustments were made, if they could be made, Canada would not be nearly so prosperous. At the same time, it does not follow that Canada has as little leverage as our political leaders sometimes pretend we have. The United States' back is against the wall on energy supply. The tar sands could be an important piece of the puzzle regarding stable future energy supplies for the United States.

Canada should openly express a willingness to produce additional tar sands oil, beyond what is already approved, *as and when that energy can be extracted in a carbon-neutral way*. We might also say in a timely and discreet way (perhaps in January or February 2009) that we would be much more comfortable with increasing oil exports over time if the United States agreed to join with us as a partner in the next round of global climate change agreements. Remembering the acid rain issue, we first must commit to coming into Kyoto compliance in very short order well this side of 2020, rather than just continuing to stall on effective action. A key point in the discussion has to be that the United States cannot possibly achieve energy security without making a fundamental commitment to the very same things that are necessary to slow climate change.

If that fundamental point becomes even more widely recognized in the United States – and many Americans, of course, understand this already – the world will be a much, much better place.

Canada Is Different

Canada is different from the United States in some quite fundamental ways. I have come to understand that in very real and personal terms since immigrating here from the United States in 1967. Canadians are not only different, but also proud of those differences. On average, Canadians are more internationalist. We tend to be less militaristic, less nationalistic, more open to the potential effectiveness of public institutions, and more concerned with the less well off among us and throughout the world.

I sensed some of those differences within weeks of arriving in this country and have only rarely had any reason to doubt them in the forty years I have happily lived here. I believe that Canada is a more caring and far less harsh place, the climate notwithstanding. Indeed, geography, climate, and the bilingual, multicultural facts of life are what have helped to create many of our social and political differences. In particular, Canada's more globalist orientation and our greater willingness to use state power to advance society and economy could yet help us take a North American lead on climate change.

It seems curious to argue that Canada is more globally oriented than the United States, the world's sole superpower, a nation that dominates the world economically, militarily, and politically. But Canada does derive a higher proportion of its economy from trade; Canadians contribute proportionally more to international causes than Americans do; and proportionally more Canadians hold passports and travel outside national borders. We watch more movies and read more books that come from outside the country. Still, these features are not the crucial difference – which is that Canadians are disinclined to view themselves or their nation as exceptional, as somehow uniquely placed or possessing qualities that render it exempt from the rules of the international order. Indeed, most Canadians would prefer a deeper, stronger, more multilateral, international order.

This condition is hardly surprising. It is in part a result of the ease with which the United States dominates Canada's most important context: North America. Canada's relative influence on events in a global setting is greater than its influence has been within North America. In a global context Canada has a wide array of natural allies with similar outlooks, both in the Commonwealth and beyond. While we see eye to eye with Mexico on some global matters, in North America our multilateralist view of the world and Canadian's historic willingness to temper economic market outcomes seem increasingly out of step.

In many ways Canada seems to be a European nation that somehow drifted to the wrong side of the Atlantic. While many Americans distrust the very idea of global governance, Canadians embrace it enthusiastically. Support for the United Nations is much higher in Canada than it is in the United States. Canadians are reluctant to go to war and often quick to embrace internationally sanctioned peacekeeping efforts. Canadians are even largely unconcerned with national defence. This is not, as some Americans might assert, because Canadians assume they will be protected by the United States, but because most of them cannot imagine that any other nation would want to attack us. Canada's geography, climate, and general demeanour in the world are, it seems, sufficient defences against an outright invasion of Canadian territory.

Terrorism, though, is a concern. Terrorism, Canadians know, could produce violent events on Canadian soil, but what follows from that for Canadians is altogether different from what seems to have followed for many Americans. For most Canadians terrorism is primarily a police and perhaps an immigration matter. The military would be called into play only in terms of a direct response – for instance, to temporarily protect the new government in Afghanistan and to thereby, at least theoretically, keep that nation from again harbouring international terrorists. Terrorism is not in any sense the basis for a national crusade, even though many Canadians were killed in New York on September 11, 2001, and a potential terrorist attack based in Toronto was thwarted in 2005. Canadians also react strongly against overly aggressive police measures, such as the 1997 RCMP pepper-spraying incident at the meeting of the Asian Pacific Economic Cooperation meeting on the University of British Columbia campus, and the even more extreme Taser shooting that led to the death of a young Polish immigrant at the Vancouver airport in November 2007.

Canadians are often strongly disinclined to play a role at the United States' side in its approach to the world (as Australia and Britain have been more willing to do). Canadians, on the whole, have neither the inclination nor the audacity to try changing how other nations approach the world. We do not instinctively distrust the motives, or doubt the capabilities, of all nations save our own. Canadians tend to see themselves as one of many, no better and no worse than others.

Americans, even many very progressive, tolerant, non-xenophobic Americans, buy into the "greatest nation in the world" mindset, in part if not in whole. More than that, they often see other places as deeply flawed, more deeply flawed than their own nation, which is somehow especially blessed. Centuries of prosperity and power have contributed to a sense of exemption and a deep distrust of any and all ceding of decisions to any international authority or body. Most Americans would only place their trust in global action on climate change if that action were initiated and led by their own country. Otherwise many see it, or will see it, as a foreign threat to their way of life.

Indeed, Americans see many things as foreign threats to their way of life. As the lead-up to the occupation of Iraq made clear, many Americans deeply distrust even Europeans, even nations with which they have been allied for centuries. Obviously many other Americans are also deeply doubtful, even embarrassed, by these tendencies, but they are nonetheless tendencies that can be played upon in times of stress. Even a mild proneness to arrogance and xenophobia, when combined with a military budget that equals the combined military budgets of the rest of the world, is worrisome. Little wonder that Canadians, while fond of the neighbours that are so like us in so many ways, are also wary.

All that said, the general absence of such tendencies in Canada combined with an internationalist outlook born of minor power status makes Canada far more inclined than is the United States to support a collective global effort to combat climate change. Canadians are more willing to adapt even though Canada's energy use is every bit as extreme as that of the United States. Accordingly, climate change denial and economic fear-mongering have proved to be a very hard sell here. Even Conservative appeals to support a made-in-Canada solution fall flat, perhaps because Canadians are just not inclined to imagine that they are exempt from either global rules or the laws of nature.

Canadians historically have also been far more willing than Americans are to use the state to bind the nation together to accomplish collective purposes. Again, this is in part a function of climate and geography. In such a large and thinly settled country, transportation and communications to the furthest reaches of the land are often uneconomic. The building of national rail, road, and air links required that public institutions play a strong role. National radio and television networks required public investment.

More than that, in a sparsely settled land economic development and economic competition were accomplished more easily through the creation of state enterprises. Most Canadian electrical utilities were created as public enterprises, and those investments were much like road and other infrastructure investments: they made it possible for private investment in the resource sector, manufacturing, and retail to flourish. Canada also historically used public corporations to increase competition within the economy, as in the competition between Air Canada (once a public enterprise) and Canadian Airlines (and other carriers), between CN Rail and CP Rail, and more recently between Petro-Canada and the several foreign-owned integrated oil companies.

In many cases these public enterprises were established by conservative governments, even Conservative governments (though not recent ones). The effort was less a matter of active ideology than a pragmatic acceptance that economic, physical, and demographic conditions at the time made such initiatives necessary. In time those initiatives became a normal part of how things were, and are, in Canada. More than that, Canada developed health-care and higher education systems that are predominantly within the public sphere, even more than is the case in some European nations, let alone the United States. Canadians are still quite comfortable with governments that are active in the economy in ways that enhance and stimulate an economy led by private enterprise.

The country also has a long tradition of research in public institutions – research that ultimately finds its way into private-sector activity. The National Research Council has played an important role in this regard, though it has more recently ceded ground to research based in and around Canada's public-sector universities and university-industry joint ventures, which often include a generous proportion of public money. It is no accident that the Blackberry and Research in Motion (RIM) arose in and around one of Canada's leading electronics- and computer-oriented research universities. The massive public-private medical research organization linked to the University of Toronto, MaRS (Medical and Related Sciences), is a contemporary example in this tradition.

Such traditions exist in other nations, including the United States, but the multisectoral co-operation in Canada runs deeper in the sense that here such organizations do not hesitate to incorporate public purposes as well as ultimately allowing private-profit opportunities to emerge. This Canadian tradition of openness to public activity within the economy should help not only to make action on climate change more effective, but also to make possible efforts that simultaneously benefit both public purposes and the Canadian economy.

Canada has not been a leader on wind energy. Denmark got into that game in a big way first and as a result has created a major industry. Other European industries have established other initiatives in terms of energy-efficient industrial equipment. Japan has set the pace on hybrids, and Mercedes-Benz has sold its energy-efficient "smart cars" all over the world, while Detroit languishes. But many, many opportunities remain – in biofuels from cellulosic sources, in cold-weather-adapted wind generators, in carbon sequestration, in rail-based public transit vehicles (in which Canada already has a solid market share), new kinds of insulation, geothermal energy extraction, and anaerobic digestion of agricultural wastes in cold climates.

The possibilities are extensive. The challenge is to get government to replace incentives to fossil fuel production with a mix of regulations regarding all energy production – and incentives to energy technologies and practices that result in GHG reductions. Those incentives could include energy taxes (in place of income or other taxes), incentives to early adopters, public institutions willing to place early orders, and public-private research funding. If that can be done, the products that result will be exportable, as will some of the fossil energy we might have consumed had we not moved ahead. Foot-dragging will see Canada importing technologies developed elsewhere.

Canada must dare to be assertive. Again, Canadians must remember the struggle regarding acid rain and look to change our own behaviour and use that change to influence other nations. Canada is a natural bridge from Europe and the rest of the world to the United States. Canada could demonstrate that big, rich nations with gluttonous energy habits can adapt. The Canadian government needs to stand up to the considerable power of our own provincial governments as well as the power of the Americans who want and expect us to maximize energy exports and are indifferent as to whether Canada can meet its international climate change obligations – especially those among them who are unwilling to accept their own moral obligations in this regard. The possibilities on that front would seem to be better with a Democratic administration in power in Washington, but it also remains to be seen whether any Canadian prime minister is ready to take such an assertive stance – some critics would read it as a threat to our well-being to not eagerly supply every drop of oil that the United States is prepared to buy – even in private.

Thinking Like Oil Sheiks

What candidate for Canadian prime minister in the coming years could think like the leader of a powerful and consequential nation committed to moving the world in a decisively different direction on climate change? The answer is none – not unless a lot of Canadians begin to think that way first.

The oil wealth of the Middle East flows into public and private treasuries at a massive rate, and those treasuries wander the world buying assets, from office towers and hotels in London and New York to significant blocks of shares on the stock exchanges of New York to the stock exchanges themselves. They also wield political influence without apology.

Canadians, even after the recent increases by the Alberta government, still offer bargain-basement royalty rates and bargain-basement tax treatment for the profits of mostly-foreign oil companies. Canada has also agreed in advance to never reduce oil exports once they are established (as in the Free Trade Agreement) and did not re-bargain that agreement when a third partner was added (making NAFTA), even though the third partner (Mexico) did not agree to those same terms with regard to its oil. Many people in Canada apparently still imagine that the United States is doing us a favour by buying our oil, as if we were selling them something that they could easily buy elsewhere or for which they could easily develop substitutes.

That day is gone. We are not selling beaver pelts. Without Canadian oil the U.S. economy would grind to a halt. Not only does the United States need Canadian oil more than ever, but multinational oil companies desperately need places to invest their massive profits in the production of future oil. Almost no other nations will let them anywhere near their oil fields, except perhaps within management or purchasing contracts on much tougher terms than Canada has thus far demanded.

Canada should not only avoid undermining its chances of complying with Kyoto in order to export tar sands oil before low-carbon technologies are in place, but should also be telling the Americans that we will only sell them oil when we can do so without compromising our global obligations – and then only on the condition that they too join in a post-Kyoto global agreement with teeth.

It is hard to imagine Stephen Harper thinking in those terms, or for that matter Stéphane Dion acting in that way. Perhaps I misjudge them. Post-Kyoto negotiations could even be Harper's Nixon-goes-to-China moment. That is, as a long-time friend of the oil industry, just as Nixon was a tried and true cold warrior, Harper could possibly bring the industry around to accepting that Canada must act effectively on this issue, just as it was said that only Nixon could make a move towards U.S. recognition of China. Perhaps that is a plausible suggestion, but the oil industry exerts a great deal more power within Canada and within the Conservative Party than Taiwan or even perhaps Cold War forces did in Nixon's America.

More likely Harper is politically astute enough to know that to do well in elections now, and for a long time to come, Conservatives must blunt climate change as a political issue. For the many things that he has said and done over the years on this issue, Harper has not had a moment of public confession or offered any explanation. He has not said, "Oh my goodness, I had no idea that this was actually a real problem. I'm really sorry. I will work hard to make up ground." In his utterly unapologetic fashion he has had little if anything to say about the new evidence coming in, let alone about having changed his mind. He merely works to stretch out the time horizon and to slowly move Canada back towards an "anti-Kyoto alliance" that includes the United States, China, India, and perhaps even Japan and Korea.

In late September 2007, shortly after speaking at a UN gathering designed to bolster Kyoto and kickstart the next round of negotiations, Harper announced that Canada would join the Asia-Pacific Partnership on climate change. This group, pushed into existence by Bush, serves as a counterweight to the pro-Kyoto group led by Germany, France, and other European nations. Including Canada, this group of nations produces half of the world's greenhouse gases. These nations, with the possible exception of Japan, want to go slow on taking further steps and above all to avoid mandatory targets on emissions. They want to keep the rules as vague as possible, indeed if possible to avoid "rules" altogether.

They want to set "goals" and leave every nation to "do their best," "in their own way." In other words, the United States and the rapidly developing giants want to do nothing that might run the risk of slowing economic growth by even a small fraction or involve changing consumption behaviours in any significant way. They are, perhaps, willing to make an effort to create and deploy new technologies as and when (and if) those technologies are ready, but not to impose burdens on their industries or their citizens. China wants to continue to open new coal-fired power plants as rapidly as possible, and the United States is unwilling to push its utilities or its auto industry hard or to establish, for example, new national supports for a rapid expansion of public transit.

The United States argues that it cannot be expected to agree to binding reductions because China and India will not (those countries have not until now produced much by way of emissions and even now produce far, far less per capita). China, for its part, may be reluctant to act as fully as it might so long as the U.S. government avoids taking strong measures regardless of what some U.S. states, cities, or corporations are prepared to do.[11] Harper covered Canada's new membership in this group by saying that he hoped to bring them into the Kyoto fold. With Harper at the helm, though, any post-Kyoto agreement would more likely be altered to accommodate the addition of these new members, and Canada's weight will be added to the side of minimalist effort.

Harper did his best in the autumn of 2007 to put Kyoto aside without explicitly defying the world and the wishes of the Canadian electorate. Mostly he did his best to pretend that Canada's signing of the Kyoto Protocol had never happened. He asserted Canada's willingness to make big cuts in the longer term, but in doing this he offered a slight of hand manoeuvre. As journalist Bill Curry pointed out, Harper was using 2006 as the baseline for cutting emissions, rather than Kyoto's 1990 baseline. Thus when Harper called for cutting emissions by 50 per cent by 2050 – as favoured by the European Union (and opposed by the United States) – the amount he had in mind "is not as onerous as the 50 per cent called for by European Union nations."[12]

"Not as onerous" is an understatement. Some 50 per cent based on 2006 is not much better than meeting what Canada agreed to in Kyoto many years ago, but it would come about thirty-eight years too late.

Deception and delay also guided Canada's performance at the Conference of the Parties meeting in Bali in December 2007. There Canada, acting as a proxy for the United States, sought above all to keep firm and specific commitments out of the conference communiqué. Canada's primary achievement in Bali was to play a role in relegating specific GHG reduction figures to a footnote and in avoiding any suggestion that wealthy nations had special obligations with regard to reductions. Given the scale of the challenge facing the world and the paucity of support for these views, it is hard to imagine a more negative role.

Somehow Canada must find a way of overcoming its internal divisions, both electoral and regional. Somehow it must remember its own history and strengths, and imagine that it is, of all things, a leading North American nation with a crucial role to play on behalf of the world.

Seven

The Challenge

Leadership and Long-Term Thinking

EFFECTIVE LEADERSHIP requires thinking ahead. Political leaders and influential nations both need to have a clear sense of where the world is going. There is no magic to this. Their guesses just need to be better than others' guesses. Effective leaders need to get themselves and their corporations, institutions, families, cities, provinces, nations, or the whole world moving in the right direction sooner than anyone else. That is the way to get ahead or to make it into the history books as something other than a scapegoat. In a globally integrated world, nations need to get on the right side of history all the more, but Canada has not assumed the leadership that so obviously stares us in the face and for which others are so clearly less suited.

For what ends is Canada now helping to put the climate of the planet at risk? Ironically, the pace of investment in a single industry in a single region of the country is driving the value of the Canadian dollar to the point that it is putting jobs and prosperity elsewhere in the nation at risk.[1] As if that were not enough, hyper-rapid development that will not wait for better extraction technologies bets wrongheadedly that the resource in question will not be worth more a few years later than it is now.

More accurately, to allow new commitments to tar sands expansion to proceed without demanding that future developments incorporate low-carbon extraction and processing technologies is to bet that in

2020 or 2025 oil is more likely to cost $40 per barrel than $140, or $240 for that matter. That is just about the most foolish bet I can imagine making. To not put too fine a point on it, Canada is throwing away its international reputation for integrity to take a bet that we are almost certain to lose. The rate of return on any investment is, of course, impossible to determine in advance, but is there any doubt that oil in the ground is *better* than money in the bank? Yet, even with the recent increase in royalties by the province of Alberta, Canada continues to subsidize the oil industry.

Writing in *The Walrus* magazine, Don Gillmor provided a summary of the level of incentive offered in the recent past: "While oil sands production increased by 74 percent between 1995 and 2002, royalties from the oil sands decreased 30 percent, a function of the low 1-percent royalty rate." Taxable profits can be avoided under these rules so long as existing operations are expanded continuously. Disputes only arise if investments are deemed new rather than expansions of existing facilities.[2] Canadian taxpayers still cover the tab on investment writeoffs, while oil companies expand their holdings and operations with money that would otherwise have ended up in the Canadian treasury, available for investment in an alternative Canadian energy future.

Even the much-touted Alberta Heritage Savings Trust Fund has not kept pace with comparable funds in Alaska or Norway. Alaska has gone further, ensuring the loyalty of all Alaskans to oil extraction by making "dividend" payments to every citizen in the $1,000 to $1,900 range. For many reasons that is a doubtful idea, but the approach is better than using the money to accelerate resource extraction to the point where little hope remains of meeting the Kyoto obligations.

Federal tax policies especially could be greatly changed in ways that would help to spread energy investments around the nation, just as changes in Alberta's royalty rates could help to diversify the Alberta economy, a long-standing provincial objective. Additional federal revenues associated with phasing out rapid oil industry investment writeoffs could provide incentives to energy options that do not result in GHG emissions. An exemption to phasing out the writeoffs could be allowed for investments that radically lowered emissions from tar sands operations.

In general, the focus of all governments needs to be rebalanced to shift from supply subsidies to demand-reduction subsidies and from supports for fossil fuel production into alternative energy options. After all, fossil fuel production hardly needs the support, to say the least. Industry protestations notwithstanding, such shifts would most likely in themselves not slow the development of the tar sands all that much, or for long. Tar sands development would remain profitable, and investment flows would continue even if they slowed modestly while adjustments were made. If any slowing of tar sands investment did occur, low-carbon extraction technologies would then have more time to catch up.

The result of altered incentives and disincentives could be greater diversification and dispersion of energy investments. The larger share of Canada's energy investments would probably still flow to one region of the country and one form of energy, but a better balance might gradually emerge. Energy efficiency investments almost inevitably occur where energy is consumed – wherever goods are produced and sold and where most Canadians reside. Alternative energy options, from ethanol to wind to small-scale hydro to nuclear power, are also widely distributed and are arrayed in geographic patterns that are very different from fossil fuel deposits.

Talking to Americans

Canada faces a profound dilemma. Given that Canadians are both habitually deferential with regard to our neighbours to the south and globally oriented, we cannot avoid offending someone in this matter. We will let down either the world at large or, at least for a time, the Americans. Canada cannot be both deferential and globally oriented when it comes to tar sands development. Canada can defer to the U.S. desire for secure energy imports, or it can do its part to ensure global climate stability. We really can't have it both ways, unless and until we find a clean way to extract tar sands energy and come to a decision about how much tar sands extraction is enough tar sands extraction.

The way out of the dilemma lies in making a factual case to the American public at large, especially those – probably a substantial majority – already concerned about climate change. Remember acid rain. We can lobby state legislatures, governors, and the Congress. Communicate with U.S. environmental organizations. Publish op eds. Explain the dilemma. We want to export energy to the United States to the extent that we reasonably can, but not to an extent or at a rate that forces us to violate our climate change obligations. We will honour existing agreements, but need to change the rules regarding future production and to gradually replace the existing and already committed production with technologies yet to be developed and proven. As our efforts unfold we could invite prominent and sympathetic Americans to visit and to then communicate to their country regarding the steps we are taking here. Perhaps an independent Canadian-sponsored documentary film – something that this country of ours is very good at creating – could graphically illustrate what is possible.

Canada could also show that North American retail businesses can change their ways – businesses that are in many cases the same companies operating on both sides of the border. Although, as George Monbiot lays out so clearly, corner stores and supermarkets are gigantic energy sinks, they do not have to be so; and many are willing to change given high energy prices. Canada could impose rules, and Canadian utilities could offer incentives that would accelerate change in that sector.

Canada could show that alternative energy production works. We could phase out coal-fired power plants (or sequester the carbon), and have our cities take effective measures to reduce sprawl. Indeed, many U.S. cities are now ahead of Canada on the issue of sprawl, and many jurisdictions are moving on alternative energy too. Chicago and Portland, for example, are building homes within their cores at a rapid rate. Chicago is putting hybrid buses into its fleet and New York is replacing its taxi fleet. What Canada could do is unify and accelerate such actions *on a national basis*, something that the United States is far from doing.

Canada has a difficult time seeing itself as a leader in, or of, North America. But in present circumstances it has the opportunity and the responsibility to lead. Canada can play a leading world role on climate change only if Canadians see the opportunity and have the courage to act. Canada cannot come close to meeting its own obligations on climate change, even on a delayed basis, without engaging in a dialogue with the United States about how much energy we are prepared to sell them, when, and under what conditions.

There is really no avoiding that discussion. The only alternative is to pretend that the link between climate change and accelerated energy exports does not exist. Real Canadian leadership would seize the opportunity, see the ways in which it connects to an emerging desire in the United States to rethink what has happened to its national reputation in this new century, and help to bring the United States back into the world system on the world's terms rather than on that country's terms.

Getting Canada out in Front

Canada could play a crucial role in the coming negotiations regarding post-Kyoto climate change initiatives. Accepting that role, however, needs to begin with a very large national *mea culpa*. It also requires taking clear and rapid steps to achieve compliance with the nation's Kyoto requirements. We need to show that reducing emissions is possible in a rich North American nation with suburbs and comfort – in effect with all the comforts of what looks to those south of the border like home. The necessary steps to achieve that compliance would be large ones, and they would almost certainly need to include: phasing out coal-fired electricity in Ontario and sequestering carbon from coal-fired plants in Alberta; a moratorium on new tar sands development commitments; strong incentives to urban commuters to switch to transit; a rapid phase-out of incandescent light bulbs; and tough new standards for appliances and automobiles.[3]

A commitment to those steps would provide an admission ticket to a significant role in what might be done globally in the years ahead. As by far the largest producer of GHGs, the United States does not need a ticket to enter the game at a high level, but Canada does. Canada, however, could have leverage over the United States on this issue at a point when a U.S. government that is not actively hostile to climate change action comes to power.

Even now signs are appearing to indicate that the internal balance in the United States may be shifting to the point at which Canada could help to move the U.S. government incrementally further. The shift in attitude within the United States is apparent on several fronts. Obviously Al Gore has single-handedly made an impression on many Americans as well as many Canadians; and many actions have been undertaken, but not widely publicized, even in jurisdictions governed at the time by Republicans – including California, Massachusetts, and New York.

Then, strikingly, in September 2007, a leading member of the 1981–89 Reagan administration also spoke out on climate change (again drawing only modest media attention, but that too could change, and Canada could help to change it). George P. Shultz, Reagan's secretary of state (1982–89), in a column in the *Washington Post*, took a stance that distanced his views from those of the Bush administration (much as other former Reagan and Bush I officials have distanced themselves from George W. Bush and other neo-conservatives).[4] Shultz noted that during his tenure global action was taken on ozone depletion through the Montreal Protocol. In his words: "The greenhouse gas problem is more broadly recognized today than it was during the Kyoto Protocol negotiations. Moreover, the protocol is running its course, so a new treaty is needed. That treaty should have a different structure – one that ultimately achieves universality."[5]

Most interestingly, in that same article Shultz had this to say about the participation of China and India:

Do not expect China, India and other developing countries to accept what amounts to a cap on economic growth. They will not – and cannot – do that. We must create market incentives for them to cut emissions while continuing to grow and find actions that are economically feasible in a relatively low-income environment. We may also need to give them extra time, even allowing them some short-term emissions growth, before requiring them to reduce their emissions. This is similar to the way we accommodated developing countries under the Montreal protocol.

These remarks suggest the beginnings of a dissolution of the impasse wherein the United States refused to participate in Kyoto on the grounds that China and India are "exempted" from stringent GHG requirements. That position of refusal is, of course, absurd. The wealthy nations of the world are responsible for more than 90 per cent of the anthropomorphic GHGs that are now in the atmosphere. An elementary sense of fairness would suggest that the non-rich countries have a right to emit an equal per capita amount before they are morally obliged to reduce their emissions. The problem with that approach, of course, is that by the time fairness is achieved and China and India begin a long course of reductions, the planet might well be uninhabitable.

Another reason for China, India, Brazil, Mexico, and other high-growth nations to resist is that without some increment of additional energy use (not necessarily, of course, fossil energy) they would find it difficult to imagine that they could continue to reduce poverty within their populations. Clearly there needs to be some middle ground. Angela Merkel, the German prime minister, has suggested that a formula be developed on *per capita* emissions that would see emissions in some nations rising and those in others falling until they met at some point within the range of the present differences among nations.

In such a scheme the greatest challenge would fall on nations such as Canada and the United States, which have the highest per capita emissions. This is clearly the time for Canada to step up and accept that challenge and to act decisively, immediately, and effectively rather than just talk about goals and obligations.

It is probable that the United States, no matter who is elected in 2008, will object to such an arrangement. The per capita differences remain very, very large, and to get to a point at which total global emissions are low enough to be unproblematic, the highest users might well need to reduce their per capita GHG outputs by 80 to 90 per cent, while Western Europe and Japan would only need to reduce emissions by half of that and many nations could double emissions and some even increase per capita emissions by tenfold. We know how Harper feels about anything like that possibility – we need only recall the fund-raising letter he wrote after becoming leader of the Canadian Alliance.[6]

Reductions on that order would be a challenge that the United States and Canada would be hard-pressed to meet within the lifetime of anyone alive today. Yet it is hard to see what other arrangement would be fair and any less difficult to achieve without a lot of people continuing to go hungry. We must also remember that fossil fuels are a finite resource and that the world must in any case soon learn how to live largely without them. What nations are better equipped to find ways of doing that than the nations of North America and Europe? This is perhaps especially true of Canada. We are less densely populated than Europe or the United States and thus have extensive opportunities for biomass, wind, and solar energy, and we have highly advanced technological capabilities that could be turned to energy efficiency and the search for new energy options.

Harper deserves credit for being unwilling to lecture other nations without Canada having done its share. But unless he explicitly rejects his earlier perspective on the subject his credibility will be limited, to say the least. As on acid rain, Canada must act first, and as in the case of upper atmosphere ozone, the concept of technology transfer is essential to global solutions. Canada should be developing carbon sequestration technologies or technologies that obviate the need for coal to sell to China – not selling China more coal and worrying about how hard done by we are in North America over the only truly sensible global approach to dealing with climate change.

The Scale of the Challenge

As nearly as we can make out, and some uncertainties remain here, the scale of the temperature shift that the planet is now almost certain to undergo as a result of greenhouse gas emissions (even if future emissions are reduced) appears likely to be as great as any shift since the end of the last ice age twelve thousand years ago.[7]

One reason for this eventuality is not widely discussed or appreciated. It is called residence time – the amount of time that carbon dioxide released into the atmosphere remains in the atmosphere. Most of what is emitted now will be in the atmosphere fifty years from now, and much of it (up to 30 per cent) will be there two hundred years from now. Some emissions will be there one hundred millennia into the future before they are incorporated into carbonate rocks. There are numerous schemes for increasing the uptake of carbon by plant life (not a high-risk effort) or the ocean (a very high-risk possibility, as are the schemes to deliberately place particulate matter into the atmosphere to offset climate warming). Setting those schemes aside – since none of them are being seriously considered – even if all GHG emissions stopped tomorrow the average temperature on the planet is already set to rise. The question is how far and how fast it will rise.

At the end of the last ice age, when the last change as large as the one we are now facing occurred, there were but millions of humans in existence and that existence was already largely migratory. As a species we humans adapted with remarkable skill and spread our numbers across the planet. Such a successful adaptation to a similar change now will not be so easy because there are billions of us, many of whom cling to life by a thread that could easily be snapped by climate change. But more than that, human adaptation today faces other challenges.

First and foremost humans now occupy every corner of the planet. There are few places to which we could migrate that are not already more or less fully (or even excessively) occupied. Our numbers are also such that we already press the capacity of the planet to produce the food that we must eat and in many places the fresh water we must drink. Most of the world's oceans, rivers, and lakes are already overfished. Humankind only just escaped from significant shortfalls in global grain production a few decades ago, and the capacity to keep pace depends on climate stability. More urgent, perhaps, drought and the loss of glacier ice threaten the potable water supply of many regions of the world – in Australia, Africa, and Asia especially.

Global food production depends on the continued use of fossil fuels for tilling, harvesting, fertilizers, pesticides, and the transportation of food from points of production to points of consumption. More than that, one of the possible replacements for fossil energy is expected to come from these same food production lands, and the capacities of those lands would be challenged by climate change. All of agriculture could function with perhaps 10 per cent of present fossil energy use, but if that overall total is to be reduced some of the reduction may need to come from agriculture. We cannot necessarily add to the total by opting to desalinate sea water or by moving fresh water over vast distances – both of which tasks require vast inputs of energy.

As important as total human numbers is one other crucial difference between the adaptation of twelve thousand years ago and today: the existence of modern weapons (including the tactics and logic of terrorism) and armies. Some twelve thousand years ago the weapon of choice was probably the club. If climate change and peak oil provoke mass migrations (and what is already going on in Darfur and arguably in Iraq is the result of climate change and peak oil), migrants may come up against the organized capacity to seize or defend territory.

All of this is a terrifying scenario, but surely – we assume – humanity will be able to mount a collective response. We assume that we are more advanced, more capable, than we were twelve thousand years ago. Still, even that seemingly obvious assumption might be in doubt. Our numbers make us far less resilient than we once were. The option of reverting to hunter-gatherer mode is lost to us in all but a few isolated locations – and those places would not remain isolated for long. More than that, much of what might be hunted and gathered will also most likely be challenged by climate change.

Nor, for all our new and improving communications capabilities, is it at all clear that, even knowing the scale of the danger, humankind is any more able to make a species-wide decision of this magnitude than we were in millennia long past. That is the primary question at hand: Is humankind capable of making a collective decision that runs even a small risk of limiting economic output in the short term to protect overall well-being in the longer term? Can we make a binding global decision to reduce global fossil energy consumption in a timely way (or to sequester greenhouse gases at great expense)? All of our efforts so far have not significantly slowed the rate of increase of fossil energy consumption, let alone stopped the increase, let alone reduced total consumption – and there is no question that these actions must be done and done quickly.

It remains an open question as to whether any nations, let alone all nations, are, or will be, able to accept slower economic growth unless or until it is forced upon them by crop failures, flooding, and/or drought. Some European nations have perhaps done it to some extent, but will anyone follow them, and are many of them prepared to see the changes through to the next level? Whether they or anyone else will be able to make the transition from the intensive use of fossil energy without a significant loss of economic output is at this point unknown.

Can nations or the whole species significantly reduce fossil fuel consumption and still grow economically? There are two different questions here, because offloading manufacturing is one way of achieving a reduction on the national level but not at the global level. Those nations that move most adeptly to reduce fossil fuel use may well be able to show economic growth, but those that are left importing both very expensive oil and the technologies necessary to use energy more efficiently may well not; nor will those nations that have a more limited technological capacity and face the worst climate change effects. It would also seem likely that any nations that become less well off than previously will be especially resistant to, or incapable of, taking costly actions to improve energy efficiency (though nothing reduces GHG emissions like a steep recession).

It also seems likely that unless *all* nations can bring themselves to act collectively, few nations indeed would be capable of effective action over an extended period. Those that continue to try to engage in that action in the absence of universal participation might well soon be overcome by the futility of their efforts in the face of the denials, refusals, and inabilities of others. More than a few nations must lead the way and demonstrate that reductions can be achieved while economic growth continues. Canada, with its resolutely internationalist history, has the aptitude and experience required to help move the world back on a path towards global governance, not "just" on climate change but also on other crucial matters.

Far and away the greatest tragedy of the twenty-first century thus far is that rather than moving towards an improving capacity for global decision-making we have plunged backwards at least a half-century in less than a decade. The United States has rejected or withdrawn from a long list of international accords and has even violated the Geneva conventions, with its attorney general pronouncing that vital treaty "quaint." The same government has consistently denigrated and undermined the capabilities of the United Nations.

This rejection of the very tentative steps that the world had previously taken towards building a capacity to make global decisions is all the more tragic in the face of the unprecedented challenges of climate change and peak oil. These problems cannot be resolved without a global co-operation that goes well beyond anything that has ever been achieved. In effect all but the very poorest nations in the world must agree sometime soon to reduce the use of fossil fuels, absolutely and immediately, or at least within the time frame of the next agreement. These steps will involve considerable economic sacrifice in the short term to avoid greater collective losses in the longer term. I do not think that there is any precedent for this anywhere in human history.

One way or another major adaptation will be necessary for us and for our children and grandchildren. Some among us will be pushed or induced to walk away from activities that we might otherwise have undertaken, or many, many others will be forced to do so at some point down the road. At present there is no decision-making capacity to make such decisions at a global level. There never has been such a capacity, and it will not become easier to achieve if the most serious possible effects of climate change or peak oil ever take hold. The only time to create a global capacity to act collectively might well soon be past.

If the wealthiest nations do not reverse the growth in fossil fuel use first, that reversal will not happen on a global scale (other than through the effects of starvation and massive economic disruption). If the most internationalist of nations do not lead the way to new capacities for global decision-making, those capacities will also not fall into place. If some nations do not push or drag the United States towards active involvement in the effort, the world will most likely not succeed in significantly slowing the course of global warming.

All things considered, perhaps no nation is as well placed as Canada to advance this effort.

Notes

One: Introduction: A Personal Reflection

1 See Al Gore, *An Inconvenient Truth* (New York: Rodale, 2006); or see the website of the Intergovernmental Panel on Climate Change <www.ipcc.ch>.

2 For an extensive discussion of the effects of climate change, see, for example, Dinyar Godrej, *The No Nonsense Guide to Climate Change* (Toronto: Between the Lines, 2006). The scientific evidence for climate change is available in thousands upon thousands of studies.

3 Until 2004 Australia was meeting its comparatively easy targets, but did not sign Kyoto. Thereafter an expanding resource sector overwhelmed the targets, but in November 2007 a government committed to climate change action was elected.

4 See Guy Dauncey, *Stormy Weather: Solutions to Global Climate Change* (Gabriola Island, B.C.: New Society Publishers, 2001).

Two: Canada, Oil, and the World's Stage

1 Such a decision is obviously fraught with all manner of political problems. I will do my best to address them as we proceed.

2 Consumption figures are from Thomas Friedman, "Doha and Dalian," *The New York Times*, Sept. 19, 2007 <www.newyorktimes.com>.

3 See Robert Paehlke, *A Tale of Three Cities (Kyoto, Baghdad, and New Orleans): American Power and Global Citizenship*, forthcoming; and Michael T. Klare, *Blood and Oil: The Dangers and Consequences of America's Growing Dependency on Imported Oil* (New York: Henry Holt, 2004).

4 See George Monbiot, *Heat: How to Stop the Planet from Burning* (Cambridge, Mass.: South End Press, 2007), p.104.

5 This option involves a double challenge: there is an energy loss in the conversion from electricity to hydrogen; and additional energy inputs are needed to move the gas over long distances. It might prove to be the case that the more economic use of remote wind energy is to use it in remote regions to power snowmobiles, lighting, heat, and other needs to avoid the cost associated with transporting gasoline from Southern refineries for generators in the North.

6 The platform of the Green Party of Canada calls for extensive investments in research regarding cellulosic ethanol.

7 For further discussion and a list of sources, see Robert Paehlke, "Environmental Sustainability and Urban Life in America," in *Environmental Policy: New Directions for the Twenty-First Century*, 6th ed., ed. Norman J. Vig and Michael E. Kraft (Washington, D.C.: Congressional Quarterly Press, 2006), pp.57–77.

8 National Roundtable on Environment and Economy (NRTEE), *Advice on a Long-Term Strategy on Energy and Climate Change in Canada*, Ottawa, June 2006.

9 See NRTEE, *Advice on a Long-Term Strategy*, p.1 (Summary of Key Findings).

10 See Green Party of Canada, *Vision Green*, Ottawa, 2007 <www.greenparty.ca>.

11 Peak oil is a controversial term suggesting that at some point very soon the available supplies of conventional oil will reach maximum output and then begin a long, slow decline. I do not pretend to know if oil extraction history will follow a bell curve pattern over its full history. I use the term, though, because I believe we should follow that pattern by choice. There is a good chance that, climate change aside, we may be nearing the point at which half of the world's existing oil has been used and therefore we should begin to slowly reduce extraction in order to reduce the risk of abrupt supply reductions at some point in the future. Indeed, prudence would suggest that we stretch out the remaining supplies as long as possible even if we are not at the point of having used half of the total that exists.

12 Jeffrey Simpson, Mark Jaccard, and Nic Rivers, *Hot Air: Meeting Canada's Climate Change Challenge* (Toronto: McClelland and Stewart, 2007), especially pp.197–245. Although the authors of this book, published in September 2007, take a different approach than I do in their analysis of the politics of climate change, many of their broad conclusions are similar to mine – which is perhaps a reflection of just

how painfully obvious Canada's flawed record has been.

13 Many Canadians appear willing to accept a carbon tax, but many do not. A Strategic Counsel poll in January 2008 found that "almost half of Canadians surveyed endorse the idea of taxes being used to bring down greenhouse gas emissions"; 49 per cent of those surveyed "said they would be willing to pay a carbon tax as part of an effort to deal with climate change." Brian Laghi, "Anxiety Grows about Economy, Jobs, Poll Finds," *The Globe and Mail*, Jan. 15, 2008, p.A4.

14 Green Party of Canada, *Vision Green*, p.35.

15 Doug Struck, "U.N. Global Warming Report Sternly Warns against In-action," *The Washington Post*, Nov. 17, 2007 <www.washingtonpost.com>.

16 That is why the solution must be global; but a global solution will not be achieved until the United States stops pointing fingers at China and others and instead accepts some reasonable solution that begins with sharp GHG reductions in North America and future obligations that are historically and scientifically informed and fair to all nations.

17 A global cap-and-trade system would need a considerable bureaucracy to administer it in order to ensure confidence, before payments were made for reductions, that significant emissions did actually occur.

Three: The Tar Sands Dilemma: Do Oil and Democracy Mix?

1 Statistics in this and the following paragraph are primarily 2007 figures from the Energy Information Administration (EIA), U.S. Dept. of Energy, Washington <www.eia.doe.gov>; or the U.S. Senate Committee on Energy and Natural Resources, Washington <www.energy.senate.gov>.

2 Govinda R. Timilsina, Nicole LeBlanc, and Thorn Walden, *Economic Impacts of Alberta's Oil Sands*, vol.1, Study no.110 (Calgary: Canadian Energy Research Institute, October 2005).

3 See Dan Woynillowicz, *Oil Sands Fever: The Environmental Implications of Canada's Oil Sands Rush* (Drayton Valley, Alta.: Pembina Institute, 2005) <www.oilsandswatch.org>.

4 Hydroelectric power involves few if any GHG emissions – unless the flooding behind dams drowns whole forests, in which case the forests will cease absorbing carbon dioxide and the decaying will cause the emission of methane.

5 The figures in this paragraph were taken from the website of the Bloc Québécois <www.blocquebecois.org>; and also discussed in Robert Paehlke and Nicola Ross, "Greening Politics: Canada's Leaders Strut their Colours on the Political Stage," *Alternatives: Canadian Environmental Ideas + Action* 33 (June 2007), pp.20–24.

6   Both figures are from the CBC News website, Nov. 25, 2005 <www.cbc
    .ca/news/background/kyoto/climatechange> (May 20, 2007).

7   Robert Sheppard, "Piping Carbon Back into the Ground," CBC News,
    March 9, 2007
    <www.cbc.ca/news/background/kyoto/capturing-carbon>.

8   See, for example, the Sierra Club website <www.sierraclub.ca/prairie/
    files/OilSandsVision&Principles.pdf>.

9   Parkland Institute, press release, March 7, 2006, Edmonton <www
    .ualberta.ca/parkland> (May 22, 2007).

10 Ibid.

11 Tar Sands Watch (Polaris Institute), "Dehcho Leader Calls for Tar
    Sands Moratorium," March 22, 2007 <www.tarsandswatch.org>.

12 Ibid.

13 See, for example, James Laxer, *Canada's Energy Crisis* (Toronto: James
    Lorimer, 1975).

14 Alberta NDP, press release, Edmonton, Feb. 5, 2007
    <www.ndpopposition.ab.ca> (May 22, 2007).

15 Kevin Taft, "2007 Speech from the Throne for an Alberta Liberal
    'Team Taft' Government," Edmonton, Feb. 20, 2007
    <www.albertaliberal.com>.

16 Woynillowicz, *Oil Sands Fever*.

17 This is acknowledged regarding cap-and-trade and coal-fired power
    plants by at least one leader in the U.S. power industry. See David
    Crane, "We're Carboholics, Make Us Stop," *The Washington Post*, Oct.
    14, 2007 <www.washingtonpost.com>.

18 I know this from personal experience, having owned, during the days
    of the first energy crisis, an 1860s stone house with most windows
    facing north into a vast sweep of open land. On the other hand, I
    would also note that Jack Layton and Olivia Chow of the NDP have
    retrofitted their old and challenging home in downtown Toronto to a
    high standard of energy efficiency. It can be done.

19 Again, I am not advocating nuclear power, merely saying that we
    should debate its selective use. Canada could do what needs to be
    done without it or with only very limited use, but densely populated
    and rapidly growing nations such as China, India, and parts of In-
    donesia may have no other option if they are to limit fossil fuel use
    and continue their essential economic growth.

20 "Harper's Green Stand over the Years," *The Globe and Mail*, Sept. 24,
    2007 <www.theglobeandmail.com>. The comment was made on Feb.
    3, 2007.

21 See, for example, the multination study by Peter Newman and Jeffrey

Kenworthy, *Sustainability and Cities* (Washington: Island Press, 1999); and David Malin Roodman, *Paying the Piper: Subsidies, Politics, and the Environment* (Washington: Worldwatch Institute, 1996).

22 The classic study on this point is Sam H. Schurr, *Energy and Economic Growth in the United States* (Washington: Resources for the Future, 1962).

23 Oil prices in early 2007 were still lower than they were in the early 1980s in terms of 1980 dollars.

24 For an extended discussion, see Paehlke, "Environmental Sustainability and Urban Life in America," pp.57–77.

25 Even if urban dwellers opt to drive, they are more likely to walk to and from parking lots, to and from lunch, and on short errands. Many studies have focused on this issue. See, for instance, "Driving Longer Means Larger Waistlines. Researchers Examine Link Between the Environment and Obesity," AP, May 31, 2004: "Spending more time behind the wheel – and less time on two feet – is adding inches to waistlines and contributing to the nation's obesity epidemic, a new study concludes." Also, Lawrence Frank et al., "Obesity Relationships with Community Design, Physical Activity, and Time Spent in Cars, *American Journal of Preventive Medicine*, June, 2004; and Michael Babad, "Researcher Links Gas Price, Obesity; Commuters More Likely to Walk, Bike or Take Transit When Costs Rise," *The Toronto Star*, Sept.14, 2007, p.B3.

26 Monbiot, *Heat*, p.192.

27 I will not extensively elaborate on issues related to energy supply beyond the earlier discussion of wind energy and the tar sands; there are many good sources on this subject. For non-technical readings that emphasize social, economic, and political factors, see, for example, Sarah James and Torbjörn Lahti, *The Natural Step for Communities* (Gabriola Island, B.C.: New Society Publishers, 2004); and Ted Nordhaus and Michael Shellenberger, *Break Through: From the Death of Environmentalism to the Politics of Possibility* (Boston: Houghton Mifflin, 2007).

28 These are 2005 figures from Government of Canada, "Top Five GHG Emitters by Province," Ottawa <www.ghgreporting.gc.ca>.

29 See, for example, Tad Patzek, "The Cellulosic Ethanol Delusion," *Energy Resources Digest*, June 28, 2007.

Four: From Toronto to Kyoto on the Ambivalence Express

1 On the early history of the "greenhouse effect," see Gale E. Christianson, *Greenhouse: The 200-Year Story of Global Warming* (Vancouver: GreyStone Books, 1999).

2 See Richard Elliot Benedick, *Ozone Diplomacy: New Directions in*

Safeguarding the Planet (Cambridge, Mass.: Harvard University Press, 1991).

3 See Robert Paehlke, "Green Politics and the Rise of the Environmental Movement," in *The Environment and Canadian Society*, ed. Thomas Fleming (Toronto: ITP Nelson, 1997).

4 See Kathryn Harrison, *Passing the Buck: Federalism and Canadian Environmental Policy* (Vancouver: UBC Press, 1996).

5 Ibid.

6 Robert Paehlke, "Spatial Proportionality: Right-Sizing Environmental Decision-Making," in *Governing the Environment: Persistent Challenges, Uncertain Innovations*, ed. Edward A. Parson (Toronto: University of Toronto Press, 2001).

7 Deborah VanNijnatten and Douglas MacDonald, "Reconciling Energy and Climate Change Policies: How Ottawa Blends," in *How Ottawa Spends 2003–2004*, ed. G. Bruce Doern (Toronto: Oxford University Press, 2003), p.77.

8 Ibid.

9 Ibid.

10 See Robert Hornung, "The Voluntary Challenge Program Will Not Work," *Policy Options*, May 1998, pp.10–13 <www.irpp.org>.

11 Sierra Club of Canada, *1997 Rio Report Card*, p.2 <www.sierraclub.ca>.

12 Ibid., p.3.

13 Don Gillmor, "Shifting Sands," *The Walrus*, April 2005 <www.walrusmagazine.com>.

14 Simpson, Jaccard, and Rivers, *Hot Air*, p.73.

15 "Not Ready for Kyoto, Chrétien Adviser Says," *The Toronto Star*, Feb. 22, 2007 <www.thestar.com>.

16 See David Suzuki Foundation website <www.davidsuzuki.org/files/CNDsolutionsNEW.pdf>.

17 Ibid. <www.davidsuzuki.org/files/Fuel_report.pdf>.

18 See Climate Action Network Canada website <www.climateactionnetwork.ca/e/resources/publications/can/kb-report-9–2002.pdf>.

19 "Harper's Green Stand over the Years," *The Globe and Mail*, Sept. 24, 2007. Harper made the statement quoted on Sept. 27, 2006.

20 John Kerry was widely accused of pandering when photos appeared of him in brand-new hunting garb and carrying a new gun during the 2004 U.S. presidential election campaign.

21 Elizabeth May, "The Kyoto Debate: Separating Rhetoric from Reality," *Policy Options*, December 2002; for this article, see <www.sierraclub.ca>.

22 Ibid.

23 Ibid.

24 For a longer discussion on this point, see Robert Paehlke, *Democracy's Dilemma: Environment, Social Equity and the Global Economy* (Cambridge, Mass.: MIT Press, 2004).

Five: Ratification and Beyond: Losing Face While Making Money

1 William (Bible Bill) Aberhart (1878–1943), a radio evangelist in Alberta in the 1930s, railed against Eastern Canadian interests and went on to be elected premier of the province in 1935.

2 See, for example, Jack Aubry, "Canadians Willing to Adapt to Halt Climate Changes," *Ottawa Citizen*, Aug. 7, 2000 <www.ottawacitizen.com>. The general support for action has continued; see Laghi, "Anxiety Grows about Economy, Jobs, Poll Finds," *The Globe and Mail*, Jan. 15, 2008, p.A4.

3 Quoted in "While Klein Attacks, Chrétien Backs Kyoto," CBC News, Oct. 24, 2002 <www.cbc.ca>.

4 Sierra Club of Canada, *2002 Rio Report Card*, p.31 <www.sierraclub.ca>.

5 The Sierra Club drew a line between Klein and Albertans on this point, saying, "Over the top fear mongering of the Alberta economy losing 'trillions of dollars' has no basis in fact, and is rejected by the Alberta public. According to recent polling, the majority of Albertans favour ratification." Sierra Club of Canada, *2002 Rio Report Card*.

6 Margaret Wente, "It's the One-Tonne Kyoto Fraud," *The Globe and Mail*, Jan. 15, 2005.

7 All quotes are from a copy of the letter published in *The Toronto Star* and other papers; see also "Harper's Letter Dismisses Kyoto as 'Socialist Scheme,'" CBC News, Jan. 30, 2007 <www.cbc.ca>.

8 I have not found survey data to support this conclusion, but I have heard anecdotal observations to this effect from many individuals from several different provinces. As well, the current and previous leaders of the Green Party of Canada have had past connections to the old Progressive Conservative Party.

9 Prime Minister Paul Martin, address at the UN Conference on Climate Change, Dec. 7, 2005 <www.pco-bcp.gc.ca>.

10 Bill Curry, "Ottawa Planning More Cuts to Climate-Change Programs," *The Globe and Mail*, Nov. 25, 2006, p.A16.

11 "Ambrose Slams Liberals at UN Climate Summit," CBC News, Nov. 15, 2006 <www.cbc.ca>.

12 Ibid.

13 Quoted in Elizabeth May, "The Clean Air Act Is Not about Cleaner Air or Action," *The Globe and Mail*, Oct. 11, 2006, p. A21.

14 Ibid.

15 Ibid.

16 Regarding Canada – U.S. policy differences, see, for example, George Hoberg, "Comparing Canadian Performance in Environmental Policy," in *Canadian Environmental Policy: Ecosystems, Politics, and Process*, ed. Robert Boardman (Toronto: Oxford University Press, 1992), pp.246–62.

17 Curry, "Ottawa Planning More Cuts to Climate-Change Programs."

18 Jeffrey Simpson, "The Prime Minister Is Blowing Hot Air," *The Globe and Mail*, Oct. 11, 2006, p.A21.

19 Ian Urquhart, "Ottawa Too Partisan on Emissions," *The Toronto Star*, Dec. 17, 2007.

20 See, for example, Chris Mooney, *The Republican War on Science* (New York: Basic Books, 2005); and for more recent examples, see *The New York Times*: Andrew C. Revkin, "Climate Change Testimony Was Edited by White House," Oct. 25, 2007, and "Memos Tell Officials How to Discuss Climate," March 8, 2007.

21 Intensity has not necessarily disappeared from governmental policy design even if it has gone out of favour within public pronouncements.

22 "Yes, Things Are Heating Up, New Environment Minister Says," CBC News, Jan. 5, 2007 <www.cbc.ca>.

23 Ibid.

24 Margaret Wente, "A Questionable Truth," *The Globe and Mail*, Jan. 27, 2007, p.F1.

25 Margaret Wente, "That Ol' Beer-Fridge Paradox," *The Globe and Mail*, Dec.4, 2007.

26 Wente, "Questionable Truth."

27 Ed Morgan, "Readers Offended by May's Analogy," Canadian Jewish Congress OP-eds, May 2, 2007 <www.cjc.ca>; also published in *National Post* (Toronto). The quotation from Stephen Harper is from Thomas Walkom, "Why the Fuss over May's Metaphor?" *The Toronto Star*, May 3, 2007.

28 See Walkom, "Why the Fuss over May's Metaphor?"; and Aaron Wherry, "Let He Who Has Not Made a Neville Chamberlain Reference Cast the First Stone," *Maclean's*, May 2, 2007 <www.macleans.ca>. See also Rick Salutin, "History Meets the New Politics," *The Globe and Mail*, May 4, 2007, p.A21.

29 Jonathan Kay, "Thomas Walkom Defends Elizabeth May's Indefensible Comment," *National Post*, May 3, 2007 <www.nationalpost.com>.

30 Matthew Bramley, "Analysis of the Government of Canada's April 2007 Greenhouse Gas Policy Announcement," May 28, 2007 <www.pembina.org>.

31 Peter Gorrie, "PM's Climate Plan 'Misleading,'" *The Toronto Star*, Sept. 22, 2007, p.1.
32 Ibid.
33 Governor General, *Strong Leadership: A Better Canada*, Speech from the Throne, Ottawa, Oct. 16, 2007, pp.14–15.
34 Ibid.
35 Gloria Galloway, "Harper Plays Down Hopes for APEC Climate Deal," *The Globe and Mail*, Sept. 7, 2007.
36 Ibid.
37 Ibid.

Six: Frozen Governance in a Melting World

1 U.S. figures can be found at U.S. Dept. of Energy, Washington <www.eia.doe.gov> and Canadian figures at Environment Canada <www.ec.gc.ca>.
2 See, for example, Pew Global Attitudes Project, "Global Unease with Major World Powers," an extended forty-seven nation survey, June 27, 2007 <pewglobal.org/reports/pdf/256.pdf>.
3 See chapter 5, p.84 here; and chapter 5, note 3.
4 Actually, "they" don't say that at all, I made it up. But maybe "they" should say it.
5 Some five hundred U.S. mayors have signed onto the U.S. Mayors Climate Protection Agreement, in stark contrast to the U.S. federal government and not in most cases, one might guess, without political intent. See the website of the city of Seattle <www.cityofseattle.net/mayor/climate>.
6 For example, on Nov. 8, 2007, the State of California sued the U.S. Environmental Protection Agency for failure to act on the state's request for a waiver on vehicle emissions standards. The state hopes to raise the standards considerably with regard to GHG emissions. If it succeeds, given its size, it could change the standards for all cars sold in North America. This is widely known in policy literature as the "California effect."
7 See Energy Council of Canada <www.energy.ca.gov/low_carbon_fuel_standard/>; Conference of New England Governors and Eastern Canadian Provinces, *Report on Climate Change Projects*, August 2002; and Apollo Alliance <www.apolloalliance.org>.
8 See "Inside Washington: Congressional Insiders' Poll," *National Journal*, April 1, 2006, pp.5–6.
9 Thomas L. Friedman, "Lead, Follow or Step Aside," *The New York Times*, Sept. 26, 2007.

10 Bill Curry, "Ottawa Signs on to Rival Emissions Pact," *The Globe and Mail*, Sept. 25, 2007.

Seven: The Challenge: Leadership and Long-Term Thinking

1 In November 2007 I visited Corner Brook, Nfld., where the only major industrial employer in the region, a pulp mill that has been in operation since the 1920s, was laying off workers because, even though it is a modern operation, it cannot compete with places in which the trees grow faster or the wages are lower. The high value of the Canadian dollar is costing jobs in places in which they are badly needed. Canadians do not need to subsidize those jobs – we need to consider the optimum level of energy exports, all things considered.

2 Gillmor, "Shifting Sands."

3 In April 2007 the federal government, following the lead of Nunavut and Ontario, announced a ban on incandescent bulbs by the end of 2012. Ontario has repeatedly delayed the promised closure of coal-fired plants, which is now scheduled for December 31, 2014, even though critics argue that the province could act well before that date.

4 Both Brent Scowcroft and Clyde Prestowitz, for example, have objected to the neo-conservative turn in U.S. foreign policy, and James Webb, a secretary of the Navy under Reagan, is now a Democratic Senator from Virginia and an anti-Iraq War critic. See Clyde Prestowitz, *Rogue Nation: American Unilateralism and the Failure of Good Intentions* (New York: Basic Books, 2003).

5 George P. Shultz, "How to Gain a Climate Consensus," *The Washington Post*, Sept. 5, 2007.

6 See p.87.

7 This perspective is well documented and was elaborated on in an article by scientist-blogger DarkSyde on the Daily Kos website, Sept. 23, 2007 <www.dailykos.com>.

Index

Marquis Book Printing Inc.

Québec, Canada
2008